*Air Crash*

# Air Crash

## THE CLUES
## IN THE WRECKAGE

by
# FRED JONES

*Foreword by* JOHN CUNNINGHAM

A COMET BOOK

A Comet Book
Published in 1986
by the Paperback Division of
W.H. Allen & Co. Plc
44 Hill Street, London W1X 8LB

First published in Great Britain by Robert Hale Ltd in 1985

Copyright © Fred Jones, 1985, 1986

Printed and bound in Great Britain by
Biddles Ltd, Guildford & King's Lynn

ISBN 0 86379 094 1

The views expressed by the author in this book are not
necessarily those of the Ministry of Defence

# Contents

# Illustrations

*Credits*

All the illustrations are Crown Copyright with the exception of the following: 2, Bristol (now closed); 9, 10, 11, Fairey (now closed); 12, 34, 37, author; 28, 29 Air Show spectator, unnamed; 40, 41, 42, 43, Japanese source. All sketches are by the author.

## DEDICATION

This book is dedicated to my son Frederick
Charles, who, like his father, has been bitten by
the aviation bug. With aviation ever being
made safer, it is hoped that he will not have to
follow in father's footsteps for the whole of his
career.

# 'Just a Touch of Magic'

The late Dr Percy Brooksbank Walker, CBE, MA, PhD., one of the country's leading aviation scientists and for many years Head of the Structures Department at the Royal Aircraft Establishment at Farnborough, of which the Accident Section was part, gave the James Clayton Lecture before the Institute of Mechanical Engineers in 1965. His subject was 'The Scientific Investigation of Aircraft Accidents'. He said that, despite the lecture title, the investigation of an accident was more of an art than a science. In dealing with the analysis of wreckage, he said, there were basic rules, which were more or less common ground for most investigators, but beyond these the action depended upon the individual. A solution had been produced for the cause of the accident in virtually all cases where substantial amounts of wreckage had been delivered to Farnborough, and he described the wreckage analysts as 'unusual people'; although their training was obviously important, more so were the inherent characteristics which could not be acquired.

'Give them a piece of wreckage and they will read it as another person would read a book. Give them a heap of wreckage and they will be kept busy and contented for weeks. Most of what is achieved is capable of a rational explanation but it is only fair to state that in this work of "wreckage analysis" there is just a touch of magic.'

# Foreword

The very high standard of air safety we all expect and accept today is the result of many years of hard won experience by the designers and operators of aircraft. Both have had to learn from unhappy experiences in the past how to improve the standards of design and operation as each new type of aircraft has been developed.

The Accidents Investigation Board (AIB) has made much use of the experience and facilities available to the Royal Aircraft Establishment (RAE) to help find the cause of many aircraft accidents, and Fred Jones has given us a fascinating picture of the work he and his colleagues in the Accidents Section at the RAE at Farnborough have been involved in during the last forty years.

*Air Crash* covers a period of extremely rapid development in aviation in which we have seen astonishing advances in performance of aircraft coupled with much greater reliability of engines and equipment used in today's civil and military aircraft. The book gives an insight into the very varied and immensely valuable work done by Fred Jones and his colleagues in unravelling the causes of accidents.

Those of us who have spent our lives flying in the aircraft industry and become involved in accident investigations very soon realized what a vital part the AIB and the Accidents Section at RAE, through its small but dedicated team, have played in pinpointing the cause of a failure in an aircraft, thus enabling the design team involved to remedy the failure and

13

help to achieve the high safety standard we all desire.

The Comet accident in early 1954 over the Mediterranean near Elba was a tremendous set back to British Aviation and particularly to those of us at Hatfield who had been involved with the Comet from its beginning.

The recovery of most of the wreckage over the next eight or nine months from the sea bed of the Mediterranean and its subsequent detailed examination and reconstruction by the RAE at Farnborough, culminating in the discovery of the origin for the pressure cabin failure and the destruction of the Comet, is a remarkable story. It was this particular accident investigation which opened my eyes to the astonishingly detailed and thorough work put in by that small number of people whose life's work has been in accident investigation.

I am glad to have the opportunity to say 'Thank You' to Fred Jones and his colleagues over the last forty years who have made flying so much safer for all of us.

Group Captain John Cunningham
CBE, DSO**, DFC*, DL, AE, FRAeS

# Acknowledgements

A book about rubbish! To the layman, the scene at an air-crash site must resemble a scrap metal yard, but to myself and others engaged on air-accident investigations, the wreckage contains a wealth of very valuable evidence, helping in the determination of the cause of the accident.

This book then is about the evidence found during wreckage examination and analysis, and has been written because, during the thirty-nine years in which I have been engaged on such work, audiences at lectures and visitors to my laboratory, including royalty, have always asked to hear more. Now, with the approval of the Ministry of Defence, this book is the answer. Primarily factual, any comments or criticisms that may appear are entirely mine.

Although this book is not written as an autobiography, I have been directly involved with every case discussed. But of course accident investigation is essentially team work, and much of that which is written here could only have been accomplished with the help of colleagues and associates at Farnborough in the many technical and scientific departments, at the Accidents Investigation Branch and in industry and the Armed Services. It would be impracticable to list everyone, and invidious to attempt selection however, with regards to the actual preparation of this book, I must especially thank Denis Goode, Chief Librarian at the Royal Aircraft Establishment, Farnborough, his wife Una, and Brian Kervail, now curator of the Museum at Farnborough, for their valued readings of the manuscript. My thanks, too, to Stephen Barlay, author of *Aircrash Detective*, who in recent years has continually urged me to put pen to paper and who gave me advice on how best to tackle the problem. Finally, and by no means least, for her patience, forbearing, support and understanding during the long hours when I was absorbed in the writing of this book, my eternal thanks to my wife, Heather.

# CHAPTER 1

# Prologue

'Truth is never more important than in aeroplane accident investigation, where half truths, the product of cursory, incomplete or ill-conceived and loose thinking, may place the lives of men and the future of aeroplanes in jeopardy.

'And so a deep look at all aspects is essential if an investigation is to succeed.'

The lecturer continued with: 'Take the modern aeroplane. It requires tens of thousands of drawings even to build it, miles of electrical cables, large areas of metal sheeting, thousands of rivets and screws, dozens of boxes of radios, electronics and other equipment, all in themselves made up of many parts. All of this material makes up a single slender, usually aesthetically pleasing form.

'Now crash that aeroplane into the ground and the product is up to half a million separate fragments, each fragment bearing boundaries not produced by man. Each fragment almost beyond recognition. That is the subject likely to confront the wreckage analyst or person who will be tasked with rebuilding, examining, analysing and pronouncing upon the wreckage, that it did, or did not, contain the cause of its own catastrophe.

'From the experience of examining many such wreckages, inevitably many lessons have been learned. Perhaps the most important has been the need to look twice – of course, not literally to look twice, although that can often be the case. But there is a need always to make more than a superficial study or assessment of a problem if the truth is to be obtained.

'It cannot be emphasized too much that the basic information derived from the wreckage must be obtained correctly, otherwise wrong ideas emerge which lead to a waste of effort and a continuance of accidents. Experience has shown that it may not be sufficient just to accept that the information obtained is correct: it must be an entity in itself, otherwise it may be likened to having obtained only a half truth.'

The foregoing is an extract from a lecture that I presented in 1968 during a visit to Canada. I had entitled that lecture 'The Need To Look Twice' because, as I said, 'It is not what we know, but what we imagine, that becomes the truth in our minds.' I had reached this conclusion as a result of many years not only of studying aeroplane wreckages but also of observing investigators and scientists at work.

Situated on the south side of the Farnborough airfield, in an area known as Berkshire Copse, hard by the Basingstoke Canal and near to where pioneer S.F. Cody prepared and flew his string-and-bamboo flying machine, is a cluster of buildings used specifically for the storage, reconstruction and examination of the wreckages of crashed aeroplanes, where an investigation has been considered necessary to determine the cause of the accident.

A visitor to those buildings, one day in early 1958, sees the floor of the largest covered with heaps of debris. A closer look reveals a lone figure, squatting Indian style, surrounded by the heaps. Occasionally he picks up a piece of twisted, torn or burnt metal, looks at it – perhaps nodding his head – and then puts it down again. The heaps of debris, amounting to many tons, are the remains of a once proud, sleek and elegant airliner – Bristol Britannia G-ANCA. It has suffered the ultimate fate and has been reduced to this ignominious spread of largely unrecognizable bits and pieces.

The lone figure, totally absorbed in his work, the nearby firing of a rocket along a test track, leaving him quite unmoved, is myself. If you were to ask what I was doing, I would have to reply, 'Getting to know my wreckage.' In some way I was seeking a closer affinity with this rubbish. 'Really,' you think, 'now we know that scientists are truly mad.'

What appears to the man in the street as heaps of rubbish and debris are to me storehouses packed with a wealth of evidence waiting to be released to enable me to determine the probable cause of the aeroplane being reduced to such a state. 'Getting to know' is a necessary step, in my opinion and experience, before embarking upon a lengthy and detailed study.

How can I possibly get to know such inanimate material as this torn, twisted and sometimes burned wood, metal and plastic? Can it really profit me to do so? And in fact why bother to look at the wreckage at all? Without that wreckage, accident investigation can sometimes become a meaningless guessing game, which on occasions simply leads to a repetition of disaster because the truth has not been discovered.

This was so in the case of the Comet accidents in 1954. First, G-ALYP crashed into the sea off Elba, and before the solution had been obtained, the aeroplanes were allowed to fly again, and another Comet G-ALYY crashed off Naples, with the loss of more lives.

Comet G-ALYP left Ciampino Airport, Rome, at 09.31 hours on 10 January 1954 on a flight to London. After taking off, the aeroplane was in contact with Ciampino control tower by radio telephone and from time to time reported its position. These reports indicated that the flight was proceeding according to the airline's flight plan, and the last of them, which was received at 09.50 hours, said the aeroplane was over the Orbetello Beacon. Another BOAC aeroplane, Argonaut G-ALHJ, received an in-flight interplane message from Comet G-ALYP at approximately 09.51 hours. That message began: 'George How Jig from George Yoke Peter, did you get my ...' and then broke off. Witnesses on the island of Elba described seeing a falling, flaming mass crash into the sea at about 10.00 hours. It appeared that something suddenly catastrophic had occurred on Comet G-ALYP, because subsequent search of the sea in the area where the witnesses had seen the splash revealed bodies and debris which could be identified as the Comet aeroplane.

Immediately after receiving the news of the accident, BOAC decided to suspend their normal Comet passenger services, for the purpose of carrying out a detailed examination of the

Comet operational fleet, in collaboration with the Air Registration Board and the manufacturers, de Havillands. To this end, a meeting was called at London Airport on 11 January 1954, attended by representatives of the airline, the Accidents Investigation Branch, the engine and airframe manufacturers and the Air Registration Board. As a result of that meeting, a committee was formed to consider what modifications were necessary before BOAC could properly seek the agreement of the Minister of Transport and Civil Aviation to the resumption of passenger services by Comet aeroplanes. The committee proceeded to consider at great length, and in great detail, what possible features, or combination of features, might have caused the accident.

As a result of all the inspections and tests which followed the meetings of the committee, about fifty modifications were made, and at the end of their work the members of the committee regarded fire as the most likely cause of the accident.

On 17 February 1954 Charles Abell, Deputy Operations Director (Eng) BOAC, forwarded to his Operations Director a report and papers showing in detail all that had been done since the Comet aeroplanes had been temporarily withdrawn from service on 11 January 1954. On 19 February 1954 the Chairman of BOAC forwarded the report and papers to the Minister of Transport and Civil Aviation, stating in his letter that, on the assumption that no further indication of the cause of the accident emerged prior to the completion of the inspection and modification work, BOAC considered that all such steps as were possible before putting the aeroplanes back into service would have been taken.

The position was also considered by the Air Registration Board. On 4 April 1954 the Minister was advised that: Although no definite reason for the accident has been established, modifications were being embodied to cover every possibility that imagination has suggested as a likely cause of the accident. When these modifications were completed and satisfactorily flight tested, the Board saw no reason why passenger services should not be resumed.

In the meantime, the Minister had asked the Air Safety Board, under the Chairmanship of Air Chief Marshal Sir

Frederick Bowhill, for advice on the resumption of the Comet passenger services. On 5 March 1954 the Minister was told that the Board had considered all the available information resulting from the recent investigations and had noted the nature and extent of the modifications planned as a result. This being so, the Board saw no justification for imposing special restrictions on Comet aeroplanes. The Board therefore recommended that Comet aeroplanes should be returned to normal operational use after the current modifications had been incorporated and the aeroplanes had been flight tested.

Acting on this advice, the Minister gave permission for flights to be resumed, and the first Comet aeroplane to resume passenger service took to the air on 23 March 1954.

I was not happy with these decisions, because I had seen and examined what pieces of Comet had been recovered up to the time of resumption of operations, and thought, 'How can people be certain that they have covered every eventuality when the crashed aeroplane has not been seen? No one knows what happened.' But mine was a lone voice in the wilderness among all the experts and doyens of aviation.

Just sixteen days after flying was resumed, on 8 April 1954, Comet G-ALYY left Ciampino Airport, Rome, at 18.32 hours, on a flight to Cairo. After taking off, the aeroplane from time to time gave its position by radio telephone to Rome Air Control at Ciampino and on the last such occasion, at about 18.57 hours, reported it was abeam Naples and climbing to 35,000 feet. This position, and those given earlier, indicated that the flight was proceeding according to BOAC flight plan. At 19.05 hours Cairo received a message from the Comet reporting its departure from Rome and giving its estimated time of arrival at Cairo. No further messages were received and all attempts to make contact failed. Bodies and wreckage were found on the surface of the sea the following day, below the approximate position of the transmission of the last message from the aeroplane.

Receiving the news of the accident, BOAC decided to suspend immediately all Comet services, and on 12 April 1954 the Minister withdrew the United Kingdom Certificate of Airworthiness from all Comet aeroplanes. The Comet was grounded.

The loss of the two Comets now presented a major problem whose solution was clearly of the greatest importance not only to the future of the Comet but also to civil air transport in this country and, indeed, perhaps throughout the world. The Minister of Supply instructed the Director of the Royal Aircraft Establishment to undertake at Farnborough a complete investigation of the whole problem presented by the accidents and to use all the resources at the disposal of the Establishment.

Immediately following the accident in January, arrangements had been made for any recovered wreckage to be sent to Farnborough, for examination by the Accidents Section, on behalf of the Accidents Investigation Branch of the Ministry of Transport and Civil Aviation. At the time the Comets resumed operations on 23 March after the deliberations of the committee, only floating debris and the rear pressure dome with other parts of the rear fuselage had been received, so very little direct material evidence from wreckage had been seen when the decision to fly again had been taken. Salvage operations had continued off Elba, but the water was too deep at the site of the second accident for salvage to be considered. Wreckage continued to be received at Farnborough from G-ALYP, and so, in parallel with all the other effort at Farnborough, the Accidents Section was able to make a reconstruction and detailed examination of the aeroplane. Throughout the summer of 1954 the work continued right round the clock. On 31 August wreckage arrived containing what proved to be the vital evidence pointing the way to the correct solution. Extensive testing and experimentation gave support to the wreckage findings.

The result of all the work at Farnborough was presented in a massive report, about four inches thick, before a Public Inquiry held at the end of 1954. The accident to G-ALYP transpired to have originated in a minute fatigue cracking of the skin over the passenger cabin, at a point where direction-finding aerials had been set into the fuselage structure.

Unfortunately, since very little wreckage had been recovered of G-ALYY off Naples, the Public Inquiry could only answer to the question as to the cause of this accident: 'The fact that this accident occurred in similar weather conditions, approximately at the same height and after the same lapse of time after take-off from Rome as that to G-ALYP, makes it at least possible that the cause was the same as in that case. The state of the bodies recovered was, as in the case of G-ALYP, consistent with the accident being due to failure of the cabin structure owing to metal fatigue.'

What an expensive lesson this had been for British aviation! But the wreckage had triumphed – that locked-up evidence had been released. My motto had been vindicated: 'Assume Nothing – just be certain.'

But what of my efforts with that vast expanse of debris in 1958? Was it worthwhile to delve deep and long? Yes it was, because the wreckage studies led to one piece of shafting, whose diameter was about that of a sixpence, and the manner of a breakage of the shaft revealed the nature of the problem to be resolved.

Britannia G-ANCA was built by the Bristol Aeroplane Company Limited for the development of the 300 series Britannia airliner. It first flew on 31 July 1956 and by 5 November 1957 had amassed 719 hours in 267 flights. On 6 November 1957 G-ANCA took off from the manufacturer's airfield at Filton at 10.07 hours on a test programme with an operating crew of four and eleven technicians on board. The aeroplane returned to Filton at approximately 11.55 hours. It made a complete circuit of the airfield, during which time the undercarriage was operated, but when it was proceeding downwind of the runway the gear was in the retracted position. The aeroplane was flying in a slightly nose-up attitude and at a speed considered to be lower than was normal. At an altitude of about 1,500 feet a turn to the left onto the base leg was initiated, but before the aeroplane had made any substantial change in direction, the right wing dipped and the aeroplane went into a right-hand turn in a very steeply banked attitude. There was a substantial recovery from this steep bank, and at about this stage the call 'Mayday, Mayday, Manoc 10' was made on the R/T. A few seconds later the aeroplane again banked steeply to the right and lost considerable height. At about 200 feet above the ground the angle of bank decreased momentarily, but when the aeroplane struck the ground, it was again steeply banked to the right. From the time it first entered the right-hand manoeuvre up to the time of impact, the aeroplane had turned through approximately 270 degrees. The impact with the ground resulted in the complete disintegration of the aeroplane, and the wreckage was spread over a wide area. All the occupants were killed instantly.

About fifty tons of wreckage resulted from the accident, and in the first instance this was delivered to a hangar at Filton where some reconstruction and a technical inspection were carried out by the Accidents Investigation Branch, then in the Ministry of Aviation. This did not reveal any pre-crash failure or defect in the structure or flying controls but did show that the flaps had been in the retracted condition and that no asymmetric flap condition had arisen to account for the aeroplane manoeuvres. It was then decided to remove the wreckage to Farnborough for further detailed examinations, in the first instance with the aeroplane structure and primary flying controls. The reconstruction of the wings showed that the aeroplane must have struck the ground, starboard wing low, then cartwheeled about that wing, finally sliding sideways over the ground, port wingtip leading and breaking up in the process. No evidence was found of any pre-crash defect or failure in the aeroplane or in its main controls. The studies had, however, presented a detailed picture of all that had happened to the many bits and pieces so that a background became available of all the damage. Thus the examination that followed, which included the electrical and auto-pilot systems, could be made fully in context and not as isolated studies of pieces removed from the wreckage.

Signals from gyro units of the auto-pilot cause servo motors for the elevators, ailerons and rudder to be energized. A system of shafts, levers and push-pull rods connects the servo motors with their respective main control circuits. Each push-pull rod incorporates a pre-loaded spring whose deflection operates an electrical cut-out switch connected to the electro-magnetic clutch of its associated servo motor. The push-pull rods are called 'force limiting links' and, as the name implies, they limit the loads in the system, thereby protecting it against mechanical damage. A shear neck, or weakened section, is machined into each of the servo-motor shafts so that, if the servo motor jams, the shaft fails and thereby frees the control circuit of the encumbrance. The detailed examinations of all this equipment showed that, whilst all breakages and damage seen could be associated with the impact of the aeroplane with the ground and its subsequent disintegration, there was just one exception. Failure of the aileron servo-motor shaft shear neck was compatible only with its having occurred before the aeroplane struck the ground. With this discovery, all effort was now concentrated on the working of the aileron auto-pilot channel.

However, at about this time in the investigation, an incident occurred to another Britannia aeroplane, G-AOVG, which landed afterwards successfully. Dangerous aileron control difficulties were encountered shortly after disengagement of the auto-pilot by means of the cut-out button on the pilot's control column. The emergency lasted several minutes and at times the combined efforts of the pilot and engineer were required to maintain lateral control. It was not until the power supply to the auto-pilot system was completely removed at source that the emergency was resolved. It seemed that here was a case remarkably like that which must have occurred on the accident aeroplane.

Further investigations now revealed that a design fault had been incorporated in the electrical circuitry of the auto-pilot/- flight system, resulting in single-pole instead of double-pole switching, in connection with the operation of the aileron servo-motor clutch. The fault was such that, in conjunction with a stray positive feed, malfunctioning of the auto-pilot could occur to a degree consistent with the conditions experienced during the emergency on G-AOVG. A stray positive feed was in fact later located on G-AOVG, due to defective soldering.

It was established that the design fault must have been present also on the accident aeroplane. Whilst it was not possible to discover evidence supportive of a stray positive feed from the wreckage of G-ANCA, the fact that the aileron servo-motor shear neck had failed in flight is indicative that the aileron channel electrical cut-out systems could not have been functioning correctly, and strongly suggests that G-ANCA must have suffered an emergency not unlike that of G-AOVG but that it occurred at too low an altitude for the crew to have time to resolve their problem.

Once again wreckage analysis had released the locked-up evidence to point the way to the solution. The appropriate shaft failure in the Britannia accident approximated in area to looking end-on at a wedding ring, and this had been discovered from fifty tons of crushed debris.

# CHAPTER 2

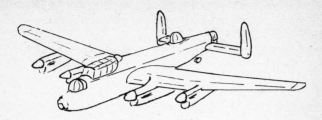

# Wartime Problems

In August 1941 I was finishing off my engineering apprenticeship at the Royal Aircraft Establishment. I was in the wind tunnels and finding the work so interesting that I felt I would like to stay in the aerodynamic business. However, fate had other things in mind for me. Our academic studies required the inclusion of laboratory work on materials and their testing, and as I had only one month of apprenticeship remaining, I was transferred to the Materials Department to make up my laboratory folder.

I was writing up some tests on specimens of plywood when Dr W.D. Douglas, the Head of the Department, walked into the room. He had a few words with my Section Leader and suddenly I found myself whisked away to assist Dr Douglas. That event was to be the start of a career which has embraced about forty years of accident investigation into everything that flies or has been shot into the sky or dropped from it, plus sundry other things as well.

Dr Douglas had been at Farnborough since the First World War, and throughout the years he had become involved in defect and accident investigations. He was always in demand by the official accident investigators of the day, but most important, he had developed techniques and know-how for analysing and considering the evidence of wreckage from crashed aeroplanes. He was laying the foundations for wreckage analysis as used today.

On that day in August 1941, Dr Douglas had a particular problem and needed someone to be his hands, to carry out his thoughts and ideas, because, of course, as a Department Head at Farnborough in wartime Britain, he was already fully occupied.

A four-engined Stirling bomber had been fitted experimentally with rocket-assisted take-off gear under its wings. Tubular cradle affairs had been fastened just below the wings, behind the engines, and a horizontal bank of solid rockets, twelve to each wing, lay flat across each cradle. The idea was to assist a heavily laden aeroplane into the air and then jettison the cradles and rockets. On test something had gone wrong. The cradles had broken free and, propelled by the rockets, had moved forwards through the lower halves of the propeller discs, then reared up and over to travel back through the top of the discs, resulting, as can be imagined, in a tremendous carve-up. Bits and pieces of tubular cradle, rocket tubing and unburnt cordite had been thrown everywhere.

My task, under the guidance of Dr Douglas, was to take all of this debris, sort it out and try to recreate the cradles and rocket tubes again, all in correct relative positions, from which an attempt might then be made to find out how and why the cradles had become detached. He told me what he wanted and then each morning we would have a debriefing and discussion on the previous day's effort. Dr Douglas explained to me what everything meant. As well as helping him, I was receiving, at first hand, a very excellent course in wreckage analysis from the most experienced tutor possible. Within only a few days I realized that I had been taken right through the exercise. Not only had I reconstructed the material, but he had led me through sequences of failures, and we were down to the probable start of the problem: a particular welded joint in the tubular structure of the cradle had been unable to withstand the forces created by the firing of the rockets. It had been a truly interesting and exciting experience.

The task was nearly finished, and my apprenticeship also within a few days of completion, when Dr Douglas asked if I would like to stay on and do this sort of work, as he thought that I seemed to have the right sort of flair and aptitude for it. I well remember saying to him, 'I suppose there will always be accidents, so I will never be out of work if I say yes.' It is not a sad commentary on aviation to say how very true that has proved, because when one considers the tremendous flying effort and passenger and freight carriage, in terms of millions of people and tons of freight, the incidence of accidents and disaster in aviation is really quite minute. But the investigative work has to be done to eradicate even that small amount. Even one accident costs lives and money.

At the time I was assisting Dr Douglas, he had just one person working full time for him on accident work, but he had been away at the scene of an accident at the time of the rocket problem – hence my introduction into the business of accident investigation. I joined forces with Keith Swainger in September 1941. He was a quiet little Australian engineer. Later he was to receive his doctorate.

The first task I had with Keith was the completion of the report on the rocket story for Dr Douglas. I shall never forget seeing that report in print, with myself as co-author. A rather proud moment: my first investigation. We then settled down to the current problem facing the accident investigators: Spitfires. Several had crashed, following structural breakage in the air, and in the laboratory around me were the remains of Spitfire wings, my first subjects for examination.

The Spitfire was an excellent aeroplane, with very responsive controls and hence was highly manoeuvrable. It was possible then for pilots to introduce high loads on the aeroplane structure if they took liberties. However, the accidents all pointed to something else causing the breakages. One interesting feature that did emerge from the various investigations was that, when the wings deflected upward under high load, a perfectly normal thing to occur, the cables that passed right across the wings and controlled the ailerons (the controls for rolling the aeroplane) became slack. The ailerons themselves were of such a design that they then altered their angles, and this in fact encouraged the aeroplane to nose up, and the wings then received an even higher loading. Thus a sort of merry-go-round situation was produced: load the wings, slacken the cables, up float the ailerons, increase the wing loading, slacken the cable and so on, until suddenly the loading was sufficient to cause failure of the wings. The remedy: pre-tighten the cables, so RAE produced a simple gadget that allowed the cables to be pre-tightened to such a degree that no amount of wing deflection in flight would cause them to slacken. Once this measure was implemented, the incidence of structural failures of this form stopped.

Incidentally, this fix for the Spitfire was then introduced on all cable controls for all aeroplanes: pre-tension cables to a standard tightness. This situation was one that I was to see repeated in different forms throughout my career. An accident would reveal a simple deficiency, not reasonably expected to have been normally considered during design and manufacture.

The pursuit of the Spitfire problem was punctuated by several 'one-off' events needing attention. For example, the Hurricane. It was of a totally different structure to the Spitfire, remarkably robust. We never saw a Hurricane that had broken up in the air, due to either structural or aerodynamic reasons. In fact the only Hurricane that I saw in pieces had been taken apart very forcibly – it had been blown up by its own weapons. In the early days of offensive rocketry, this particular aeroplane had been fitted with underwing racks and carried a battery of solid propellant rockets, each fitted with either explosive or armour-piercing warheads – in effect, artillery shells on the end of rocket tubes.

Our subject Hurricane flew to the south coast for firing trials. All went well and the aeroplane was returning. There were a few unused rockets still on the racks. Suddenly the Hurricane simply fell apart and plunged to the ground, where a mighty explosion finished the job. The aeroplane was totally destroyed and reduced to thousands of small pieces. I had to try to rebuild the wing structure and assemble the remains of the racks and the rockets to see if we had any hopes of making an examination to determine the reason for the disaster. Some tubular scaffolding provided me with the main bones, and I was able to dress it up sufficiently to give a semblance of a Hurricane wing, racks and rockets.

A slog and some patience prevailed and we found that we had a fault developed during the firing trials such that on the return flight vibrations had caused the premature firing of one rocket and the detonation of its warhead, before clearing the aeroplane. Once again the wreckage had given up its secret, and something had been learned to make the installation safer for further trials on another aeroplane. Hurricanes and their offensive rocketry went on to be a successful ground-attack combination.

Whilst searching through that terribly shattered wreckage I had been confronted with a sample of human tragedy – a deformed denture plate and a shrunken left shoe, the results of the ground fire after the aeroplane had crashed.

Keith Swainger stayed in the Accidents Section for about fifteen months after I had joined. Staff moves then became the order of the day. Dr J.B.B. Owen (Bryn) took over the Section, which was then expanded and totalled eight people at a peak period during the War. It was apparent that much work was to be expected at Farnborough. Indeed, the Section was to become the centre of accident investigations where structural failures of aeroplanes in the air were involved.

I was next to examine the remains of the prototype of the Fairey Firefly, a tactical reconnaissance machine for carrier operations with the Royal Navy. This particular aeroplane had broken up on a test flight and was being flown by a well-known pilot, Chris Staniland. The outer wings and parts of the tail had become detached, and the aeroplane had then descended in a very flat spinning attitude, such that it had landed flat on the ground without any forward motion and everywhere collapsed vertically. That wreckage told in no uncertain manner how the aeroplane had descended and impacted.

Our examinations were to reveal the manner of the disintegration and, in consequence, modifications and local redesigning were to help this become a successful machine. Equipped with special trailing edge flaps on the wings, it performed especially well at low speed, an asset for carrier work. Over nine hundred Fireflies were to see service, and some of these were to be the first Fleet Air Arm aeroplanes to fly over Japanese territory and the first to fly over Tokyo. Here was an example where wreckage analysis helped to unravel a problem in the early days of an aeroplane's development, and the type then went on to be successful. Years later I was to look back with quiet satisfaction at the small part I had been able to play in the build-up of our wartime aviation forces, when in

31

retrospect I could see other types making their marks in the battle.

Hotspur gliders started then to have a series of accidents. These were light, all-wood monoplanes with elliptical cross-section fuselages or bodies, and were built to carry troops into positions on the battlefield. The idea was to deposit a section of eight to ten men together where required, as parachuting was tending to scatter the soldiers, with all the problems that that provided. Gliders were generally towed over the landing zone at a medium altitude, released and then descended in a prescribed pattern of manoeuvres to bring them conveniently into the zone as though making a normal landing. Throughout this whole descent, the gliders would be exposed to attack from the ground by enemy fire, and their helpless occupants terribly vulnerable.

One idea was taken up to shorten the time of exposure: cast off from the tow, make a steep, high-speed dive, pull out just above the ground, and land. Trials were made to see if this would work. It did. The gliders certainly descended quickly, but it was in effecting the pull-outs that the troubles began. Some were pulled up too early and stalled and crashed; the others were pulled up too late, landing 'beneath the ground', and there were some that rounded out at the right altitude but then suffered problems due to the loadings imposed on the glider. Tailplanes broke off.

Some of the trials took place at nearby Blackbushe, for a spell a sort of satellite to Farnborough. I visited the field more than once to see broken Hotspurs. Examinations of the broken tailplanes brought to light the cause of the problem: the top boom of the tailplane mainspar was collapsing under compression loading during the manoeuvres in flight and was then exploited during the landing round-out, causing the boom to separate and the tailplane to become detached. Strengthening modifications were introduced, handling techniques altered, and the gliders became successful.

During the latter part of 1942 there arose a problem that, unfortunately, was never satisfactorily resolved throughout the war, although palliatives from time to time seemed to effect temporary cures.

Three Typhoon aeroplanes crashed following structural

32

failure in the air in which the tail units became detached. The break-up of the three aeroplanes seemed very similar, and this suggested similar circumstances, similar loadings and a possible early solution to the problem. The three detached tail ends came to Farnborough, were set up in a line in Dr Douglas's main laboratory and became the subjects of close scrutiny by many people, including myself. Our detailed studies revealed the general manner in which all three tail units had become detached. It was rather complex, with the only common feature being a predominance of upward loading on the tail at the time of separation. No ready solution was provided from our analyses and, in parallel to the wreckage studies, many mechanical and structural tests were made on the Typhoon aeroplane, by both the manufacturer and the RAE at Farnborough. These did little to resolve the problem either. The original strength tests made on a Typhoon fuselage long before the first accident had occurred had shown that the fuselage was well up to design standards and requirements.

During the next three years we were to study over twenty more wreckages, still without a firm solution as to why the failures were occurring. Each wreckage would throw up some feature which could be considered a possible contributory factor to the cause, and where possible, modifications and alterations were made to eradicate that feature from other machines. For example, tests done on a Typhoon in connection with vibrations and flutter had shown that the elevator mass balance weight could pick up violent motions and could then 'waggle' sideways. We found that several of the mass balances in the accidents had indeed failed and the fractured edges of the balance arms contained the characteristic evidence of fatigue. Arms were therefore reinforced, but the accidents still continued; indeed, there were cases of aeroplanes breaking up even with the reinforced arm.

It appeared from all the exhaustive studies, testing, calculations and enquiries that whatever happened in flight did so with little or no warning. No defects could be linked up to the failures, which further pointed to the fact that, when trouble arose, it was catastrophic.

All tests and calculations had eliminated steady loads in flight as a cause of failure, and it appeared that loads of a

33

dynamic nature were responsible. The reason for the Typhoon tail unit failures remained one of the great mysteries of the war. Fortunately the incidence of the accidents became very infrequent, suggesting that somewhere the efforts to resolve the problem were being effective, although we never knew what or why. Wreckage analysis had played its part in showing *how* the failures had occurred, but could not assist in the *why*.

Another year of war had passed and we were into 1943. During the next twelve months I was to be concerned with fourteen major investigations, nine of which involved the breaking-up of the aeroplane in the air. By now our techniques and methods had become established such that all technical problems arising in aviation accidents were being referred to Farnborough. The wreckages included such aeroplanes as Typhoons, Spitfires, Wellingtons, a Halifax, Lancasters and two aeroplanes from our American allies, a Thunderbolt and a Marauder.

The Lancaster, a four-engined twin-finned Avro bomber, did much of the offensive work against Germany and proved to be a real thoroughbred but, like most successful machinery, had teething troubles in the early days.

The earliest structural failures in Lancasters were attributed to fins – in one case an aeroplane landed with one fin intact, having lost the other in flight. The problems had arisen because the standard evasive manoeuvre used to keep the aeroplanes out of the searchlight beams, or away from fighter attacks, tended to put considerable loads on the fins. The aeroplanes would be 'jinked' or swung from side to side in violent attitudes. Modifications introduced considerable strengthening of the fins and their attachments, and after this no clear-cut fin primary failures were discovered.

Accidents did occur still, however, and in general fell into two types: those in which the wings failed under upload, and those in which the wings failed under download. The upload failures were fairly straightforward to investigate and usually resulted from high 'g' loadings in violent manoeuvres. The download wing failures were by far the more common and were nearly always associated with the aeroplane entering into a fast dive.

Although we at Farnborough had played a part in the

investigations into the Lancaster accidents, it was not until the latter part of 1943 that we had the opportunity to study three Lancaster wreckages in great detail. In all three cases the wings had failed under download, and the fins, tailplanes, rudders and elevators had also failed in the air. It seemed probable that high-speed dives were the primary reason for the failures. We knew that the fast dives were the results of the known tendency for Lancasters to nose down at high speeds. I dealt with one of the three wreckages to come to Farnborough in 1943 and have selected to describe what I did, as an example of the studies that we were making in the Section at that ime.

On 5 September 1943, Lancaster R5492, a Mark 1 machine of No. 1661 Conversion Unit, Royal Air Force, was engaged on searchlight co-operation and night fighter affiliation. At 22.38 hours four to six searchlights caught the aeroplane south-east of Exeter and the pilot took violent evasive action, performing vertical S-turns, with the nose falling in each turn. At about 15,000 feet altitude the Lancaster went into a long, steep dive, and the searchlight crews held their beams on it until they realized that the aeroplane was not apparently pulling out of the dive, when they doused their lights. The dive continued with the engine note in crescendo. There were no signs of lights or fire in the air. The aeroplane fell inverted and caught fire at ground impact. Portions of it detached before impact and made a trail of debris about 500 yards long. The shortness of the trail of wreckage indicated detachment of the pieces late in the dive, for a fairly strong wind was blowing, which, had the disintegration occurred high up, would have dispersed the detached portions over a much longer trail.

The aeroplane had a total flying time of just over 600 hours, and the pilot had accumulated about 370 hours, of which 34 had been on Lancasters, all the crew were killed in the impact. The investigators found that the outer wings and tips, elevators, fins and rudders, the starboard tailplane and the extreme rear end of the fuselage, with the port tailplane attached, had all become detached in the air. The main wreckage was largely destroyed by fire.

I received all the detached pieces and first of all reconstructed the various components. The outer wings and their tips were assembled on trestles and table tops, as were the starboard tailplane and elevator. (I had made up previously a form of table top especially for wreckage work. It was a large rectangular frame of welded angle, in steel, and a meshing of thin wire at about six-inch intervals was spread over the frame. Thus material could be supported and also examined on top and underneath.) The starboard tailplane and elevator were assembled alongside the rear fuselage portion, which together with the port tailplane now made up a complete tail unit. The fins and rudders were suspended on wires and ropes from the laboratory roof structure so that they hung vertically in position at the respective ends of the tailplanes.

I had gone to the trouble of making this rather elaborate reconstruction because there were many benefits to be gained. For example, when pieces were correctly aligned, it would be immediately apparent that skin wrinkles or other damage features were continuous or not continuous across adjoining pieces, the former indicating the presence of the damage before the separation of the pieces, the latter otherwise. The three-dimensional assembly would also enable me to appreciate the manner in which pieces may have moved to strike each other. In the case before me I could readily see how the starboard fin had failed and had struck the starboard tailplane. If I had left all the wreckage on the floor for examination, I would have been performing mental acrobatics, adding up all the evidence.

I proceeded through the wreckage with my detailed studies, noting every scratch and tear, every failure and bend, in fact everything that had happened. That wreckage had been in effect a perfect structure with clean, unblemished surfaces before the accident, so everything I saw had occurred during the accident, and if I could identify, understand and arrange all I saw in the correct order, I could re-enact what had happened to that wreckage and could therefore determine the starting feature. From that would stem the analysis to determine the 'whys and wherefores' of the event.

In dealing with the wreckage of Lancaster R5492, my examination had shown that both port and starboard outer wings and their tips had failed and disrupted in a very symmetrical manner. Both wings had failed under downloads. The ailerons had broken on both sides of the aeroplane, in consequence of the disruption of adjacent wing structure, but I could see that, before the ailerons had broken up, their upper surfaces, made up of fabric sheeting, had burst open and the free edges were teased and fluffy. This told me that the fabric edges had been exposed to the airstream for a fairly long time – far longer and at higher speeds than they would have experienced when falling free after the wings and ailerons had broken up. This was one of the useful tips that we had learned when examining fabric-covered surfaces.

This 'teased' fabric evidence suggested to me that the aeroplane had been in a high-speed dive where high-suction loads would have caused the fabric to burst and then flap and flutter in the airstream. We had seen this very feature on some Lancasters that had entered high-speed dives but recovered successfully.

The fuselage had failed just forward of the tailplane location, and here I had to examine lots of broken edges of this light alloy sheeting, which had formed the shell of the fuselage. The broken ends of the stiffening pieces through the structure also gave me valuable clues; when I had added together all the evidence of the failures around the separated end of the fuselage, I could see that the rear portion had pulled and torn away in a manner to suggest that failure had followed and was almost certainly consequential upon the failure and detachment of the starboard tailplane and elevator, leaving a one-sided tail assembly.

I now turned my attention to the starboard tailplane and elevator. The tailplane was rather like a long, shallow box, made up of four stiff booms, two at the front and two at the rear, forming the top and bottom booms of the front and rear spars. Stiffening ribs were set at intervals along the tailplane, and the whole was covered underneath and over the top with thin metal sheeting.

The starboard tailplane had broken into many pieces as well as becoming detached from the fuselage. I could see that the upper skinning had become detached from the spar booms by shearing of the rivets which attached the skin to the booms. I could also see that the rivet shear failures formed a particular pattern which was continuous along the spar boom, even over areas where boom and skin failure had occurred. This was

37

saying that the skin separation had occurred first. I could see why the skin had been disturbed and caused to become detached. It had been struck by the starboard fin, and the fin had collapsed inwards and downwards onto the tailplane. There was an excellent impression of the outline of the fin on the assembled pieces of top skinning.

My studies at the rear of the aeroplane led me to conclude that the initial feature had been the inward folding of the starboard fin. The subsequent loss of the starboard tailplane would have caused the aeroplane to nose down, and heavy downloads would then have been imposed upon the wings, leading to the detachment of the outer portions and tips in the manner that I had observed.

Why had the starboard fin collapsed? After all, both fins had been modified and reinforced, as a result of the earlier problems on Lancaster fins. My detail examination was to reveal all – but first a few words about the local construction of tailplane and fin where they are assembled to enable the evidence I found to be appreciated.

I have already described how the tailplane was a long, shallow box, with the booms along the four edges. The fin had two similar spars or posts that stood vertical, and the side skinning and a nose shape were riveted to them, again making a shallow box. The tailplane spars extended out just beyond the outermost tailplane rib, to provide two vertical faces, against which the fin posts, suitably spaced, just fitted. The fin and tailplane spars were then bolted together. The booms for the fin spars were made of angle section. Thus to make a spar, two lengths of boom were placed vertically side by side, about five inches apart, with a flange of each pointing towards the other. They were then connected by a thin sheet web riveted to the flanges. The other flanges of the booms pointed along the sides of the fin and provided the surfaces for attachment of the fin skins. Again rivets were used.

The starboard fin had become detached from its tailplane by failure of the fin posts level with the upper surface of the tailplane. The outer booms had failed first, and the fin had then folded top inwards, bending the inner booms. An interesting feature was revealed when I examined the broken end of the outer boom of the rear post of the fin. It had broken, perhaps not unexpectedly, through a section containing holes for the rivets attaching the skinning, but one of the holes had been wrongly positioned so that the drill had not simply passed through the flange but had drilled edgewise into the thickness of the other flange. The drilling had removed over a quarter of the metal through that section of the flange. Under the microscope, I could also see that cracking had developed from this drill hole in both flanges, and that cracking said just one thing – fatigue. Thus the fin post had been considerably weakened, in fact more than countering the reinforcement that had been introduced to make the fins stronger.

It seemed to me that this particular accident was probably another of the earlier dive and fin failure events, ironically on a strengthened fin – that was not!

# CHAPTER 3

# 1944, A Full and Momentous Year

Problems started early in 1944. Just four days into the New Year an aeroplane took off from Farnborough on an altitude performance climb to 20,000 feet, to be continued to 30,000 feet, exploring engine-stalling characteristics. The flight was scheduled to last about an hour, but barely a quarter of an hour after leaving the aerodrome it crashed and the pilot was killed.

The aeroplane was one of the F9/40 prototypes of the Gloster Meteor twin-engined jet plane. This particular machine was fitted with axial flow engines developed jointly by Metropolitan Vickers and the Royal Aircraft Establishment. Both aeroplane and engine were still in their early development stages. The pilot was Squadron Leader W.D.B.S. Davie, an Establishment pilot, and he was the first man to be killed in a turbine-powered aeroplane.

Squadron Leader Davie had made the first jet-propelled flight from Farnborough in the Gloster E28/39 aeroplane fitted with a Whittle W2B engine. Later in 1943 he had had to abandon the aeroplane at 36,000 feet when he encountered aileron-control difficulties. His parachute descent was probably one of the longest on record, the emergency supply for oxygen being ineffective, because his mask was torn away during the abandonment. Flight Lieutenant Davie, as he then was, managed to find the broken oxygen pipe and put it in his mouth. He dropped freely and retained consciousness until he reached a lower altitude and then deployed his parachute successfully. During the free fall he had lost his boots and sustained severe frostbite to his feet.

We received the wreckage of the E28/39 to examine, and I spent many hours in a cold chamber making tests on the aeroplane wing and aileron control circuit, at extremely low temperatures, something like minus 56° centigrade. The reduction in temperature in the chamber took many hours to achieve, and I had, in consequence, to attend at odd hours during the day and night, to make tests at the required temperatures. I had to wear special flying clothing for the tests, with heating circuits built in, flying boots and several pairs of gloves.

The tests included putting weights on a scale pan to measure the friction in the aileron circuit. The weights were about saucer size, of cast iron and of various thicknesses. Some fell off the scale pan onto the floor about two feet below. Two just shattered and two more remained intact. Without thinking, I stooped to pick them up, failed, because of the gloves, and immediately took off a glove. I promptly dropped the weight, skinning my fingers in the process. My skin had stuck to the cold metal – I had learned a lesson.

The tests proved very enlightening. At very low temperatures, a known physical phenomenon was well demonstrated. Different metals have different expansion and contraction rates with the rise or fall in temperature. Steel and aluminium, or its light alloy derivatives, are two extreme cases. The wing structure and its spars of our aeroplane were of light alloy, and the aileron circuit rods, chains sprockets etc. were of steel. At the very low temperatures, the circuits were slackening and, when operated, were jamming over sprockets and fairleads. No wonder the pilot had experienced control problems.

Back to the F9/40. It had taken off from Farnborough during lunchtime on Tuesday 4 January 1944, and employees returning to work just before two o'clock saw the silvery machine high overhead. Suddenly it fell apart and dropped to the ground in many pieces. The main portion landed just outside the Establishment on waste land. The tail unit dropped onto the Establishment foundry roof, and the remainder of the aeroplane and pieces of its port engine were distributed for thirteen miles to the south of Farnborough. The pilot had abandoned the aeroplane, but his parachute was not deployed. Squadron Leader Davie fell through the roof of a small building just to the south of the famous Black Sheds at Farnborough.

Despite the scattered nature of the wreckage trail, only the port engine and nacelle forward of the wing front spar, the pilot and his cockpit canopy, the fuselage side magazine panels and the complete tail unit were missing from the main wreckage. The extensive trail had been produced by the very strong winds of 85–95 m.p.h. in the upper air, carrying the light pieces of aeroplane and engine, which had disintegrated, away from the area of break-up. Indeed calculations and plots made for the trajectories of the falling pieces showed that, from the area high over Farnborough, the distribution across the countryside to Seale, near the Hog's Back, Surrey, was logical and fully understandable and due to the wind conditions. The wreckage was collected together and became a subject for our examination in the Accidents Section.

A straightforward story emerged from our efforts with the wreckage. The port engine had disintegrated in flight, creating an emergency which prompted the pilot to abandon the aeroplane. During his abandonment, without of course the benefit of an ejector seat, the pilot struck the aeroplane tailplane, resulting in severe mutual damage. Subsequent to this impact, the complete tail unit became detached.

Our task had been to assemble the bits and pieces, examine them and determine what had happened to that engine, how the pilot had not been able to abandon cleanly, and why the tail unit had become detached. Whilst this work was going on, we also made calculations and plots for the trajectories of the falling pieces. Apart from the early work by Dr W.D. Douglas, on falling pieces of aeroplane, the Section was in the forefront of such work.

It could be shown by simple calculation that, in general, pieces breaking away from the aeroplane moving at speed are quickly brought almost to rest with respect to the wind. Further, such parts quickly reach a falling velocity which approaches their terminal or maximum velocity of descent, which is low. Then, as the pieces fall, they drift along with the wind, and on this basis the trajectories or paths of the pieces can be plotted backward and upward from the points on the ground where they are known to have landed. No great accuracy could really be claimed for these trajectories, but by following pieces back along their trajectories and knowing that they must have become detached from the aeroplane in a reasonable order, it is possible to assign areas of the sky in which certain groups of failures or events occur.

In the case of our F9/40, it could be shown that the port engine disintegrated in an area centred roughly at 20,000 feet altitude, shedding its cowling and intake grid, which together enclose the engine normally on the aeroplane. From another area, centred around 16,000 feet altitude, came parts of the cockpit canopy, the pilot's body, his arm and some of his clothing. At a lower altitude, at about 10,000 feet, came parts of the rear fuselage and tail-unit control surfaces. A rough possible crash path was also obtained by linking these areas.

Our examinations of the wreckage were made to determine not only how things had happened but also whether any structure, control or system was defective. We found nothing in the airframe or controls which could in any way have led to the accident.

The fact that the pilot had struck the port tailplane during his abandonment also indicated that the tailplane had to be there to be struck – logical? That is how the wreckage analyst works. The trajectory plots had also placed the tail-unit detachment late in the overall event. Nevertheless, we made a thorough detailed study of the pieces involves in the separation, to determine precisely how and why the tail was removed. This was done to provide the aeroplane designer with information, because here had been a free-strength test of part of his aeroplane.

In fact, all work in this area suggested that the strength of the tail unit was adequate to meet loads which should arise in all normal flying but that in the accident situation certain other loads had been encountered, which, whilst unusual and rare, had removed that tail. We were prompted to advise the designer to strengthen the area in a particular manner. The

43

accident had brought to light a situation not normally considered in the design of an aeroplane.

The pilot's abandonment presented an interesting story. This accident had, of course, occurred before fitment of ejector seats, and abandonment involved the pilot removing the cockpit canopy, or hood, and then undoing his straps and climbing out of the cockpit. In this case the aeroplane would have been at a higher airspeed than that at which most simple abandonments had previously occurred. Additionally, the canopy was heavier, more robust and more complicated in its mechanism for attachment and detachment than such aeroplanes as Spitfires. In the Spitfire, the pilot simply slid the canopy aft along its rails, and he was clear to leave. In the F9/40, it required simultaneous pulling of two handles, one on each side of the cockpit, to cause complete detachment of the canopy for abandonment purposes. Aerodynamic loads were then supposed to lift the canopy up and away from the cockpit.

Our studies of the canopy and its mechanism showed that jettisoning had not been made cleanly, and it seemed likely that the pilot had encountered difficulty, resorting to pushing the canopy with his hand. From all our investigations, involving also a study of the pilot's clothing and plotting all that was found along the wreckage trail, it became apparent to us that the pilot must have lost his left arm and glove in the process of getting rid of the canopy. His left arm was broken and then pulled off. His overall was smeared with paint which must have come from impact with the tail unit, and there was evidence on the front of the port tailplane of such an impact. Thus, when the pilot finally fell clear of the aeroplane, it was highly unlikely that he was in a fit state to initiate deployment of his parachute. And so he fell to his death, suffering further multiple injuries as he smashed through the roof of the building at Farnborough.

The port engine had disintegrated and we were faced with a jigsaw problem with the compressor rotor, or drum that revolved. In fact, pieces of this rotor were found scattered along those thirteen miles of countryside. Since it became imperative to study every part of this engine, search parties were organized among the staff of the Establishment and despatched at the weekends across the country, along the line of the trail, to search for any missing pieces. This task, although formidable, was eased somewhat in that we quickly discovered that, although the wreckage had been distributed over such a long distance, it was all contained within a narrow strip of only four hundred yards width, along that straight path. The trail led through fields and military barracks, through built-up areas and a military sewerage works. Months after the accident pieces of the engine were being retrieved from that works when the various filter beds were dug and ploughed over. In fact, the jigsaw was finally completed with one piece, just a year later.

Once the rotor had been pieced together, the fracture edges were examined under the microscope and sequences made to determine the starting point of the disintegration. The nature of the disintegration suggested an outward explosive type separation, such as could occur if the rotor oversped. Tests on the material strength of the rotor indicated that a gross overspeed, however, would be required to cause failure. Our detailed examination, though, revealed intercrystalline weakness in one area of failure and overall low elongation properties of the rotor material, thus a reduction in the overall failing loads could well have occurred, and a relatively low overspeeding only would then have been necessary to cause the disaster.

Having reached this stage of the investigation, we now passed the problem over to the engine experts to delve further. This was in the early days of jet-engine experience, and it was obviously very desirable to find out everything about the event.

Whilst I was tidying up after the investigation, another problem had already arisen. This was to be the first of a series of five accidents during the year, each one contributing a little to the story that was to emerge.

On 26 January 1944 Stirling EH 933, a four-engined bomber of No. 1660 Conversion Unit, RAF, at Swinderby, took off on a searchlight/fighter affiliation flight, combined with a navigational exercise. The flight was to be made at 18–19,000 feet altitude. About $2\frac{1}{4}$ hours after take-off, EH 933 was heard in the vicinity of Winsford, ten miles east of the required track for the fourth leg of the exercise, between Sidmouth and Ilfracombe. It was heading west, visibility was bad, and no sign of the aeroplane was apparent. It was then heard to turn south-east, with engines spluttering, and flew towards an aerial lighthouse beacon down a valley, losing height. The engine noise stopped and four seconds later the aeroplane crashed and burst into flames. The time was 00.02 hours, and the place was Bridgetown, Exton, Devon.

An Accidents Branch inspector made the initial investigations for the RAF and found that the main wreckage was lying inverted and had hit the ground at a rather flat angle. A trail of disintegrated portions extended eastwards for $3\frac{1}{2}$ miles. Included among the detached portions were the rear fuselage, the port and starboard tailplanes, rudder, starboard wingtip, starboard aileron, undercarriage fairing and doors and the dinghy and its cover. The pilot had flown about 250 hours total but had made only one cross-country flight in a Stirling aeroplane, of duration of $5\frac{1}{2}$ hours. At the time of the accident the aeroplane had flown a grand total of 71 hours 50 minutes.

Once it was established that the aeroplane must have broken up in the air, arrangements were made for the wreckage to be delivered to Farnborough, and it fell to me to make the detailed examination. First I made calculations and plots of the paths of the detached pieces in the air. I was able to determine that the rear fuselage had broken away at about 14,000 feet, that the tailplane, elevators and rudder had become detached from the rear fuselage at 10–11,000 feet, that the parts of the starboard wing and aileron and dinghy and its cover had broken off at around 8,000 feet and other items at altitudes down to 2,000 feet. The wreckage was incomplete towards the rear of the fuselage, and unfortunately I could not identify the starting region for the fuselage separation. My detailed studies suggested that the rear fuselage had probably broken away from the main aeroplane downwards, and in consequence the parts of the starboard wing broke away downwards and the dinghy released. One intriguing problem now arose because everything pointed to the fuselage breaking before the tailplanes became detached, yet, according to the manufacturer's strength calculations, the fuselage should have been far stronger than the tailplanes. Considering the circumstances of the accident, the aeroplane was briefed for a cross-country flight at 18,000 feet, yet my investigations were pointing to failure at 14,000 feet under loading conditions indicative of a dive.

The probable reason for the accident then seemed to lie in the cause of the dive. The weather conditions on the flight route were difficult for flying, especially by a pilot with very little experience on the Stirling aeroplane, so it was possible that in thick cloud the pilot could have lost control, the aeroplane nosed over and, in the ensuing dive, excessive speed and conditions were achieved which brought about the failure of the rear fuselage. This was as far as it was possible to go in this particular case.

During May two more Stirlings crashed following structural failure in the air. The first, EE 956 of No. 1661 Conversion Unit, was on a cross-country exercise on 17 May 1944 and crashed at Ironstone Quarry, Rothwell, Northants. The second, LK 517 of No. 1654 Conversion Unit, crashed between Middridge and Shildon, Durham, on 31 May 1944, also whilst on exercise. Both aeroplanes appeared to have lost control in cloud and emerged in pieces.

The wreckage of both these aeroplanes was examined at RAF Maintenance Units, no wreckage being delivered to Farnborough. Overall disintegration of both aeroplanes was remarkably similar and also very like EH 933. In all three cases, however, I was unable to locate the origin of separation of the rear fuselage from the aeroplane. Each wreckage in turn provided local detail evidence which pointed out the general nature of the failure.

Our war efforts continued unabated. New problems were cropping up all the time as new aeroplane types entered the arena, or old ones were used in new ways, introducing new conditions not previously encountered or catered for. The long hours of peering through microscopes, handling wreckage and treading through the mud of the fields were all worthwhile, because, at the end, some simple modification to the aeroplane, or change in the manner of its operation, would ensure that a particular type of accident would not be repeated. Not only was I deeply engrossed in all that I had to do, and thoroughly enjoying it, but in some small way, too, I felt I was making a contribution to the war effort. Then came a change of emphasis. Still a contribution to the cause, but with a difference …

It was Tuesday 13 June 1944, and in the early hours four flying bombs, the German V1, crossed the coast of south-east England, on their way to London. As they did so, members of the Royal Observer Corps, who had been briefed to expect and recognize pilotless aeroplanes, detected the intruders. The bombs were codenamed 'Diver' and with the transmission of this word from the coastal observer posts, the authorities and defences became alert to a new danger from the Continent.

The V1s flew on over Kent and Sussex, awakening the sleeping people down below with their characteristic 'racketing noise'. The first to cross the coast dived towards the ground over Swanscombe, near Gravesend, and crashed at 4.18 a.m. Two minutes later, the second bomb came down at Lizbrooks Farm, Cuckfield, Sussex, killing several animals and creating minor damage. Number three reached Sevenoaks and fell at Crouch, and the fourth bomb smashed into a railway bridge, killing six people, injuring nine and destroying the bridge at Bethnal Green. News bulletins next morning reported: 'A single enemy raider was shot down over the London area', but many people in south-east England felt that something else must have happened – those strange noises were certainly not the sounds that we had all grown used to over southern England in recent years.

I arrived at the laboratory at 8.30 a.m. to start yet another new task. I spent the next few hours delving into the mysteries of the V1 bomb. The remains of the Cuckfield missile had already been delivered to Farnborough. By a quarter past three in the afternoon, my task was well in hand.

At the same time as I worked, an event of far-reaching significance was taking place elsewhere. Some debris fell out of the sky in the vicinity of Backebo, Sweden. It lay spread over an area $1\frac{1}{4}$ miles square. According to reports from the local population, a very powerful explosion must have occurred in the air at an altitude of at least 9,000 feet. One farmer claimed he was partially stunned and his horses were forced down on their knees. A few seconds later, after he recovered, he looked

48

up and saw a mass of glittering silver objects raining down from the sky. A crater fifteen feet across and six feet deep was produced by some of the falling debris. About $2\frac{1}{2}$ tons of material was collected and taken to Stockholm and stored in the Aeronautical Technical Experimental Establishment, where, on 14 July 1944, identification and reconstruction were commenced under the charge of Professor Boested. On 29/30 July the debris was suddenly removed from the Establishment, with no hint to the staff as to what was happening.

Just before lunchtime at the Royal Aircraft Establishment at Farnborough, on 31 July 1944, I was waiting along with two colleagues of the Accidents Section and several senior members of the Establishment. The previous afternoon we had hurriedly cleared an area of our laboratory and screened it off. We were now in that area. A Halifax aeroplane had landed on the aerodrome, with twelve large wooden crates. The debris which had so mysteriously and suddenly disappeared from Stockholm, equally had suddenly and mysteriously reappeared – at Farnborough. We quickly learned that it was to be our task to sift, sort and reconstruct whatever it was that was contained within those crates – of course there were no drawings or photographs – at least, initially.

The Establishment was working a 'one day off in eight' routine in those days, but for us it suddenly became 'round the clock' working. Our task had to be completed as soon as possible – we had to discover what we had received, what its performance and capabilities were, and what counter-measures might be used against it. We had received a present from Adolf Hitler, albeit through the back door. The urgency arose because we might well receive further examples, more openly and very much in anger. Would we be able to stop them?

In the Accidents Section, our main task was to try to rebuild the main structure, or airframe, whilst colleagues elsewhere at Farnborough dealt with a power plant and control systems. Initially the problem was more in the nature of an archaeological than a technical one.

We began our jigsaw. There were hundreds of pieces of mild steel sheeting – these proved to be the outer shell of our subject. We then found lengths of curved steel angles which

49

could be formed into circular frames. These appeared to be assembly joints, enabling easy assembly or breakdown, possibly for transport purposes. Other heavy steel channels were pieced together to form skeletons of four huge separate and identical fin-like structures. It was now becoming apparent that we had some kind of projectile, similar in shape to a bomb but very large, with four huge stabilizing fins at the aft end, and a large, single, venturi-shaped combustion chamber also at the rear.

A large quantity of pieces of light alloy sheeting was next considered. We divided it into two thicknesses. These must have been the glittering silver objects seen by the farmer over Sweden. We built two large jigsaw puzzles, each covering about 250 square feet, about the size of a large sitting-room. We could see now that all this material formed two large cylindrical tanks about five feet in diameter, of which one, with the thinner sheeting, tapered down at one end, to about four feet diameter. It seemed to us that these must be fuel tanks.

Pieces of light alloy casting matched together and formed a ring which was located on the aft end of the combustion unit. This ring supported four carbon vanes or controllers, which protruded into the unit. Chains, sprockets and torque shafts formed operating gear driving small tabs at the rear outer corners of the fins.

Our work continued. We were still ahead of the game; nothing unexpected had arrived over Britain to date. Soon it was possible to build our sub-assemblies together, and we found that we now had a rocket – a large cylindrical body tapering down to a near point at the front, and slightly tapering at the rear, ending with the orifice of the combustion unit. The four large fins were symmetrically disposed around the rear end and protruded beyond the orifice.

I personally then built up what was obviously warhead material and, by jigsaw and matching together and aligning the pieces, found that the warhead was conical and just fitted to the extreme front of our assembly.

The general build-up now showed how the whole weapon was compartmented along its length. First the warhead, then a section which probably housed control equipment, radio etc. Next came a bay at least twenty feet long, housing those two large tanks. This bay was about $5\frac{1}{2}$ feet in diameter. There followed a framework of steel tubing, about seven feet long, supporting pumps and a large turbine unit. The main combustion chamber was attached to the rear of the framework. The fins extended $2\frac{1}{2}$ feet beyond the rear of the chamber, making up a total length of about forty-five feet.

By now the power plant was being fully understood, and attention was being directed towards the performance of the rocket. What fuels were used? What thrust developed?

A violet stain, caused by methyl violet, was found on pieces of the forward tank, in gaskets in one of the pumps and around the rear end of the combustion chamber. Alcohol was regarded as the most likely solvent; petrol was impossible. Several points were then established to support the view that liquid oxygen was used as the oxidant.

First, a small set screw in the smaller pump had fractured, giving a crystalline break, indicative of fracture at low temperature. Low temperatures were also indicated by the large clearances allowed in the pump bearings and also by asbestos-lagged electrical cables.

The turbine assembly consisted of the turbine housing with two pumps already mentioned, mounted on either side. The turbine itself must be steam-driven and the characteristics of burners employed, and the discovery of permanganate in the system, suggested that hydrogen peroxide was the fuel used. The use of bituminous paint on a lemon-shaped tank was consistent with this.

Thus we had our main fuels, alcohol and liquid oxygen, pumped from the tanks to the combustion chamber by a steam-driven turbine unit. The performance of the propulsion system was calculated from consideration of the many features and factors now becoming available. It was interesting to note that the steam turbine gave an operating speed of 5,000 revolutions per minute, when developing 650 horsepower.

A gross thrust of 60,000 pounds seemed likely, and at about this time a document captured in France (by the not infrequent incursions by Commandos and others) referred to strength testing of the tubular framework supporting the turbine and pump assembly and gave a check on this thrust figure, because the frame had to withstand a proof loading of 64,000 pounds.

51

Using the volumes of oxygen and alcohol contained in the tanks, the duration of thrust to 'all burnt' worked out to be about 75 seconds. Weighing samples of structure and taking into account warhead filling and fuel load, gave a total all-up weight of about 30,300 pounds, or $13\frac{1}{2}$ tons.

Calculations of the flight of the rocket by our Aerodynamics Department colleagues produced a powered phase up to the height of 22 miles, and to achieve maximum horizontal range with the rocket at 45° attitude at this height, we found that maximum height turned out to be 60 miles, and the maximum range 120 miles. To achieve this, a speed of 3,500 m.p.h. was determined for the instant of 'all burnt' fuel, and this was confirmed from the physical evidence we found of kinetic heating effects on structure, paint etc. on the pieces we examined. Impact speed with the ground at the end of the flight was adjudged to be about 2,000 m.p.h.

For those of us associated with the task of determining what this thing was, and did, these final performance figures were disquieting and very frightening. These rockets could arrive on a target completely ahead of any sound – like an artillery barrage without the gunfire to warn of an impending shell.

Why had we been so disturbed and frightened by our discoveries? Because we had been assembling an example of Adolf Hitler's latest reprisal weapon, the A4 rocket, better known as V2 (*Vergeltungswaffen 2*). Now the reason for our urgent approach to the problem becomes apparent.

Our efforts were successful, because on 19 August 1944, just twenty days after work commenced, a report of the findings was produced – and the first V2 was not to fall on Britain until twenty days later, on 8 September 1944.

In that short time, the only immediate counter-measures possible were already getting under way: the identification, attack and destruction of launching sites, production plants and factories. Assuming that London was to be the main target of a German onslaught by rocket, it was possible to locate launch areas and sites on the Continent from the performance calculations. As to finding the factories and the production plants, the German engineers were very thorough: every item on the rocket carried stampings of drawing numbers, inspection markings, firm's identification, serial numbers and so on, and it was my task to note this information, pass it to the Intelligence Services and await results – 'Last night our bombers raided ...'.

Later, when the rockets were falling on Britain, all the remains were brought to us at Farnborough, where all the markings were monitored to determine rates of production and, for example, time between manufacture and rocket launching. In time we received a complete V2, which made very interesting comparison with our version.

Incidentally, we had received this debris from Sweden because of a technical problem the Germans experienced – air bursts of the rockets during the descent, but this particular rocket had only arrived over Sweden initially because of a launch and flight incident.

The background to the loss to the Germans of our 'Swedish Rocket' was later given by Major General Walter Dornberger, the man in command of the operational development of the rocket, in his book *V2*.

The development and testing work was being made at Peenemünde in north-west Germany. In June 1944 Dornberger received a call from Hitler's headquarters enquiring if any 'A4' launchings had taken place in the last few days. He checked with Peenemünde and was told that nothing had been fired. Someone must have launched something, however, because of what had happened over southern Sweden, and all indications were that an A4 had disintegrated in the air. Dornberger again phoned Peenemünde and was then told that a missile had been launched, but not to any distance. They had merely been testing the remote-control equipment for the big Wasserfal anti-aircraft rocket and had mounted it in an A4, and the projectile had gone astray. Close inquiry revealed that, while the rocket was still travelling slowly after launch, the control engineer had changed direction by eye and lost contact with it when it unexpectedly moved sideways into low cloud. The engineer had tried to bring the rocket back but evidently failed because of cloud cover. The rocket had continued flying – to the north – which, unluckily for the Germans, but not for us, took it to southern Sweden. It had been fully tanked up, so that thrust had lasted until the propellants were completely exhausted.

Dornberger reported all of this to Hitler, and was then asked if any conclusions might be drawn from the pieces of rocket in Sweden. He was able to say that, whilst the rocket itself might be understood, the Wasserfal equipment, which of course had nothing to do with the operation of the rocket on this occasion, could only pose a difficult problem to any enemy intelligence. This was in the early days following the event, and so the Germans were only concerned with the Swedes. This same situation applied in fact later, because the Germans never knew that the pieces were transferred to Farnborough.

Dornberger was not to know how close he was to the truth. Whilst the radio experts at Farnborough were able to understand fully the workings of the radio equipment from Sweden, and even determine that it must be for some sort of radio control, they were completely baffled as to its role on the rocket, which, of course, was nil; the rocket had only been the means of carrying the equipment aloft.

After this exciting and somewhat nerve-racking spell of work, life returned to normal. I recall attending a meeting in London during September which was concerned with an interesting technical investigation, and one that dramatically demonstrated the properties of a gyroscope.

Every schoolboy will have experienced the effect of holding a gyroscopic top and, when tilting it, felt the top move away at right angles, known as precessing. An aeroplane's propeller, when rotating at high speed, can be likened to a solid wheel and will possess the same properties as that top. Each blade is continually changing position around the disc and only momentarily experiences the effects of the shaft being tilted as the aeroplane's nose rises or falls or turns sideways. Nevertheless, each blade will be subjected to fore and aft bending forces at appropriate points around the disc when the aeroplane is manoeuvred. All of these effects are taken care of in the design and construction, and no problem normally arises. For very high-performance propeller-driven aeroplanes, to give smoother and more efficient performance, to enable smaller propellers to be used and to cancel out torque reaction, designers equip their aeroplanes with counter-rotating, or contra-rotating, propellers. This simply means two propellers, one behind the other, driven by the same engine but in opposite directions.

One such aeroplane was Spitfire JL349, which crashed after breaking up in flight. Among the scattered pieces of aeroplane were six broken propeller blades, all very badly shattered. The blades had been made of layers of wood, like plywood, sheathed in a material to give a hard and smooth surface. After the accident, the blades resembled matchwood. The investigators had no problem in their study and analysis of the aeroplane itself, but the wood was a different matter. Could we do something with it at Farnborough?

We painstakingly sorted out the debris and then began a three-dimensional jigsaw puzzle to reconstruct the six blades. I say three-dimensional because those layers of wood had separated, so it was a matter of making up the blade thickness in local assemblies before putting these together. Eventually enough was reconstructed to enable the mode of failure of each blade to be determined.

One of the questions we had been posed was, 'Has the propeller broken up in flight and caused the aeroplane to crash, or had the reverse happened?' Our reconstruction and examination produced a very interesting picture. The blades had all broken by bending fore and aft. Blades on opposite sides of each disc had bent in opposite directions. Because there were two sets of blades rotating in opposite directions, and placed close together in tandem, the blades had intermeshed with each other. We could see that the failures must have resulted from loadings caused by the aeroplane manoeuvring and not by a local feature on any blade.

The Stirling story now continued. On 12 September 1944 Stirling LK 499 of No. 1653 Conversion Union was carrying out a night exercise not dissimilar to that of EH 933 but over a different route. The aeroplane left its base at Chedburgh at 21.15 hours and was to fly from base to Rugby and Bristol at 12,000 feet altitude and then at 8,000 feet for the remainder of the flight. Its estimated arrival time over Plymouth was 22.43 hours. The pilot's flying experience on Stirling aeroplanes consisted of a total of 18 hours 35 minutes, dual and solo, day and night, of which he had flown one night cross-country flight of 4 hours 55 minutes, the evening before.

After taking off, the aeroplane proceeded on course and on time until it was heard just north of Plymouth, flying at approximately 5,000 feet (searchlight witness-estimated height). Shortly afterwards, at 22.45 hours, a flash on the ground was observed in the direction of Leemoor, Devon, and proved to be LK 499, crashing, following structural failure in the air. Prior to the crash, the aeroplane had been flown for a total of 276 hours.

My services were requested for the examination of the wreckage. The scene at the main site had shown that the aeroplane had struck the ground at a very steep angle and at high speed, the engines being buried fifteen to twenty feet into the ground. The trail of portions detached in the air extended approximately 900 yards to the west of the main wreckage, with small fabric portions being found up to 2,700 yards away.

On this occasion the wreckage was not brought to Farnborough, but I made a visit to Taunton, to examine the pieces at No. 67 Maintenance Unit, Royal Air Force. Upon my return to Farnborough I made calculations and plots of the falling detached pieces.

Unlike EH 933, this present case included detachment of both port and starboard outer wings. I was able to show that the rear fuselage had become detached from the main aeroplane under download, above an altitude of 3,500 feet; the tailplanes and elevators had broken away from the rear fuselage at 3,000 feet; and the wings and ailerons had left the main aeroplane at 2,000 feet. The tailplane/elevators had broken away upwards, and the wings under downloads. Once again there was a situation similar to EH 933 where the aeroplane had been briefed to fly at one altitude, yet had broken up at a lower level. Again, although the aeroplane had failed structurally, the probable reason for the accident was in the cause of the dive. I was unable, once again through lack of relevant wreckage, to locate the precise origin of the rear fuselage separation.

The weather on this occasion was good, with moderate wind conditions, and the pilot should not have experienced any difficulty in maintaining control, but his flying experience was not of a high standard, and it is possible that the aeroplane got momentarily out of control and dived to a high speed. Thus far we had repeated the accident to EH 933, but the problem remained that the rear fuselage had again preceded the tailplanes in order of failure, although the latter, on calculation, were the stronger.

Whilst I had been making my examination of LK 499, the final subject for this short series of accidents to Stirlings had taken place. This time the aeroplane had not been engaged on a hazardous night cross-country flight with a relatively inexperienced crew. The aeroplane was LK 207, of 161 Squadron, Royal Air Force, but whereas EH 933 and LK 499 were Mark 3 aeroplanes, Lk 207 was a Mark 4, the main difference being an absence of front turret and mid upper turret on the Mark 4. The pilot of LK 207 had accumulated a grand total of 382 flying hours, which included 162 hours solo, on Stirling and Halifax aeroplanes at night.

The aeroplane had flown a total of 84.10 hours prior to the accident. On its penultimate flight, of 2 hours 20 minutes, on 14 October 1944, it made a very heavy landing, following which it was placed unserviceable. The pilot also reported a sluggish artificial horizon instrument. The instrument was taken out of the aeroplane, tested and found serviceable in every respect. On 19 October 1944 the pilot did not know a new artificial horizon had been fitted, because he had signed for the aeroplane on a newly opened maintenance form, which had not carried forward reference to the instrument.

On 19 October 1944 LK 207 took off from Tempsford on a local air test of 10.59 hours, to keep the pilot in practice and to ensure that the aeroplane was operationally serviceable. Details of the flight were left to the pilot's discretion. Nothing was heard of the aeroplane until ten minutes after take-off, when it was heard making an unusual noise, and reportedly seen at about 2,000 feet, below cloud. Witnesses varied in their description of altitude and speed – from normal flight to high-speed dive. But they did agree that the whole tail end of the aeroplane 'blew to pieces' and that the aeroplane spiralled or dived down with pieces detaching. All the occupants were killed. The aeroplane had crashed at Potton, Bedfordshire. The weather on the accident flight was 8/10 cloud at 1,500 feet, 9/10 at 4,000 feet, visibility 8 miles and freezing level 5,000 feet.

The main wreckage was found lying inverted, having struck the ground in a flat attitude, with little forward speed. Detached pieces of the aeroplane were found eastwards from the main wreckage up to two miles away. The detached portions included both port and starboard outer wings, the rear fuselage and tailplanes and elevators.

The wreckage was delivered to Farnborough, and I was in business yet again on a Stirling. As on previous occasions, I was able to plot out the trajectories for the detached pieces, as well as make my detailed examination. I found that the rear fuselage had broken away from the main aeroplane, downwards, at about 5,000 feet. The tailplanes had left the rear fuselage at about 3,500 feet. Again, as in the previous cases, wing detachments had occurred under download, a prerequisite for which is a 'bunt' or nose-over, associated with a prior tail detachment. Evidence of fuselage failure before tailplanes was again established from the wreckage examination. My detailed examination of the reconstructed fuselage enabled me to trace tension failures around the top of the fuselage just aft of the mid upper turret position, and the flooring had collapsed in compression, thus indicating a downwards failure. It was interesting to see that a failure in the same region had occurred under download in a fuselage strength test made at Farnborough many years earlier for design approval purposes.

Enough evidence had been accumulated from the examinations of the first four wreckages to suggest a possible solution for the structural failures. The wreckage of LK 207 contained sufficient evidence to confirm the hypothesis.

My investigations had established the nature and location of the primary structural failure and showed that the tail of the aeroplane had been subjected to severe loads downwards, suggesting a high-speed dive. Flight tests were made and showed that, if the Stirling aeroplane was dived beyond the limiting diving speed, a nose-heavy tendency progressively built up and the aeroplane might dive away at excessive speed. As I have already indicated, the real reason for the accidents lay in finding the cause of the high-speed dives.

The accidents could all have been initiated by loss of control in difficult flying circumstances, and the structural failures were consequences. Interestingly, we had no more Stirling structural failures – of course the tide of war had swung in our favour, and Stirlings were being phased out of operations and crew training. Thus a smaller flying effort was being made, with correspondingly less likelihood of such accidents.

I often wondered, though, about the Stirlings lost on operations. I had dealt with all known structural failures to Stirlings over Britain – just five machines. 1,759 Stirling bombers had been built. They had dropped over 27,000 tons of bombs in nearly 18,500 sorties and, according to official statistics, for the loss of 641 aeroplanes. I rather doubted that the Germans could claim to have accounted for all those Stirlings.

However, the Stirling story was now closed. Some interesting structural problems had arisen but it was now too late in the life of the aeroplane to make modifications or alterations.

# CHAPTER 4

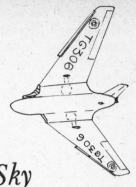

# A New Shape in the Sky

The advent of the jet engine gave aeroplane designers the opportunity to express themselves with more aesthetically pleasing shapes than had previously been possible. For example, no longer was a propeller necessary, and in consequence, undercarriages could be made shorter and the housings for these could also be smaller, thereby leading to thinner and smoother wings. New wing shapes appeared, to take advantage of the higher speeds that were now possible with the new breed of power units.

The De Havilland Aircraft Company, in the business of manufacturing aeroplanes from the early days of aviation, had always managed to produce machines of pleasing lines, and this continued to manifest itself in the aeroplanes of the new jet era.

First had been the Vampire, its short, streamlined, bomblike body set into a wing with two tail booms extended aft from the wing, on either side, to support a tailplane to the rear of the body. Perhaps a logical step forward from this was an aeroplane with a similar body but with wings that were swept back and without tail booms or tailplane – just a vertical fin and rudder at the rear of the body, and swept back too. Type numbered the De Havilland DH 108, the aeroplane was popularly and aptly dubbed 'Swallow'. Not only did it look sleek and fast, but its flight testing quickly showed that it had a remarkable performance, so much so that true level speeds were being attained considerably in excess of the World Speed Record of 616 m.p.h. then currently held by Group Captain Donaldson, RAF, in a Meteor twin-engined jet fighter.

The pilot of the Swallow during these early development flights was Geoffrey de Havilland, famous son of a famous father. In 1946 he was thirty-six years of age and Chief Test Pilot of the De Havilland Aircraft Company.

Two Swallows had been built, one capable of high-speed flight and one specifically for low-speed flight exploration. One of the high-speed versions, used by Geoffrey de Havilland, was identified as TG 306.

It was perhaps inevitable that an attempt would be made on the World Record, and TG 306 was therefore being prepared with this in mind. On the last preparatory flight, Geoffrey's intention was to embrace all the conditions which he might encounter on the actual record course, including flying at still higher speed than he had already achieved. During this flight, to be made on 27 September 1946, he was going to dive the aeroplane at something under 10,000 feet altitude, at a high Mach number (that is, at a speed closely approaching the speed of sound at that height) so as to check its controllability in these conditions. Having completed the dive, Geoffrey was going to fly level at high power, near the sea, to check the speeds and behaviour in record-attempt conditions.

On the day in question, Geoffrey flew the aeroplane off from Hatfield, De Havilland's airfield in Hertfordshire, at 17.26 hours, with full fuel. The weather over the Thames estuary at this time was 8 miles or over visibility, about half cloud cover with base at 3,000 feet and tops at 5,000 feet; there was slight gustiness, and the wind was generally from the south-south-west, at 7 knots at the surface, increasing to 25 knots at 10,000 feet.

Some time later, as the aeroplane was travelling at high speed from west to east at medium altitude, it broke up without apparently any change in attitude. Witnesses' statements were conflicting, but two wings and the fuselage were identified while parts were falling.

Being on a firm's development flight programme, the Swallow had been fitted with an automatic observer. In 1946 this sort of flight-data recorder was essentially a second panel of instruments, showing height, speed, engine revolutions and so on, facing a cine-camera; in this way a continuous record of particular parameters could be obtained. The automatic observer was salvaged from the aeroplane wreckage, which had fallen into the waters of the Thames estuary. The firm's analysis of the film record suggested that the aeroplane had broken up in a dive, probably at about 7,000 feet, at an equivalent airspeed of 580 m.p.h. and a Mach number of 0.872. This was in fact one of the higher combinations of Mach number and equivalent airspeed attained during the flight series.

The wreckage had been scattered over an area approximately 4,000 feet east-west, by 2,000 feet, north to south. The distribution of items within this area was not thought to be reliable, because it was in a tidal portion of the estuary, and recovery and pinpointing the positions of pieces of wreckage presented great difficulty to the personnel concerned. There was thus no immediate and clear-cut picture of which piece could have fallen from the aeroplane first, during the break-up.

The early stages of the investigation into the accident were dealt with by the Accidents Investigation Branch, at that time with the Ministry of Civil Aviation, and was made in collaboration with De Havillands. Because of the unconventional form of the aeroplane, and the possible repercussions that the structural failure might have on the design of future high-speed aeroplanes, the Accident Section at Farnborough was asked to examine the wreckage, with the object of determining the probable sequence of failure, when the aeroplane disintegrated.

Despite the problems of salvage, about ninety per cent of the aeroplane was recovered and taken to Hatfield, where it was re-assembled in a screened-off area of a flight hangar. Conveniently nearby was one other Swallow, and of course, being at the design firm's facilities, drawings and so on were readily available.

In seventeen days of studying the wreckage of the Swallow, sufficient evidence had been obtained to determine that the aeroplane had broken under downloads – that is, the wings bending downwards about their attachments to the body. We were able to arrange the evidence into an overall sequence of occurrence to show that the primary separation occurred in the starboard wing attachment where the root joint member had been pulled out of the centre section main spar upper boom. All other failures and detachments could be shown to be either consequential or subsequent to that of the wing failure. There was no evidence of fatigue, or any other pre-crash defect or failure; neither was there evidence of flutter in the wreckage – indeed, calculations by those expert in these matters showed that flutter would not be expected below a speed of well over a thousand miles an hour.

Disaster had struck suddenly and unexpectedly, such that Geoffrey de Havilland would have had no opportunity to jettison his cockpit canopy, to enable him to abandon the aeroplane. The canopy had broken away from the aeroplane in the air, but only because the body of the aeroplane itself had broken into pieces. The jettison hand control was still in the locked position. A strip examination of the engine had not revealed any evidence of pre-crash mechanical defect or failure and there was evidence that the engine had still been rotating at impact, but not at full power, and was probably windmilling or slowing down after its fuel feed had been severed during the break-up of the aeroplane.

Reference to wind-tunnel tests on a model of the Swallow, made at the equivalent of high-speed flight, indicated that three would be a reduction in the longitudinal stability of the Swallow, with a marked tendency to move nose downward, above a Mach number of 0.86. Above this Mach number, too, the tests showed that there were indications of control problems arising, such that, coupled with the nose-down movement, the catastrophic failure of the aeroplane, as we had determined, could be fully and completely explained. As already noted, the automatic observer information had pointed to a probable failure at Mach number 0.872 – that is, above the critical speeds obtained from the wind-tunnel tests. Flight tests had given a slight indication of a change in characteristic but not sufficient to suggest that there could be any catastrophic consequences.

It seemed that Geoffrey de Havilland had been yet another victim of man's probing into the unknown. Inevitably,

63

progress in aviation is often at the expense of life or suffering, but unfortunately most advances can only ultimately be made by piloted aeroplanes.

Reconstruction of Comet Airliner G-ALYP which crashed off Elba in 1954.

Accident site of Britannia Airliner G-ANCA, at Downend near Bristol, 1957.

Britannia wreckage at Farnborough, 1958. This is about half the wreckage.

*Opposite above*: First rocket-assisted take-off gear: port carrier from Stirling bomber. Author's first wreckage, 1941.

*Opposite below*: First rocket-assisted take-off gear: starboard carrier from Stirling bomber.

Reconstruction of Britannia wreckage in progress: wings, controls and flaps.

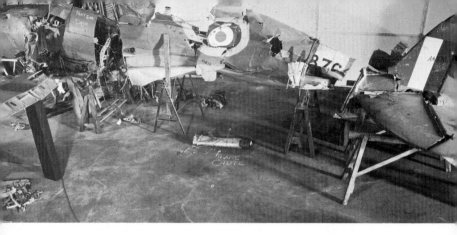

Author's first reconstruction of a Spitfire (AA876, 1941)

'Spanner in the works' – a box spanner discovered by author during strip examination of a Spitfire wing. It had been trapped ever since the wing was made.

(*Top*) Firefly Z1827 crashed in a flat spin after structural failure in the air. Note vertical collapse everywhere. View from rear of aeroplane. (*Middle*) Firefly Z1827, view from forward. Note the engine buried flat and deep. (*Bottom*) Site after Firefly was lifted leaving impression of aeroplane planform (view from rear). Deepest impression made by engine.

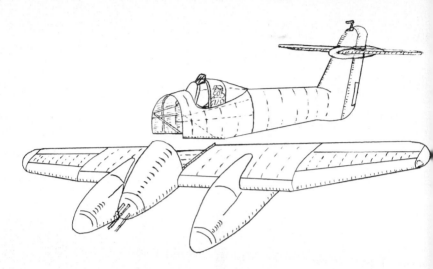

Accident to Whirlwind fighter P7103, 1942. The wings and fuselage separated in the air as shown. The fuselage dived dart-like to ground, and the nose was concertinaed to half its length by the impact.

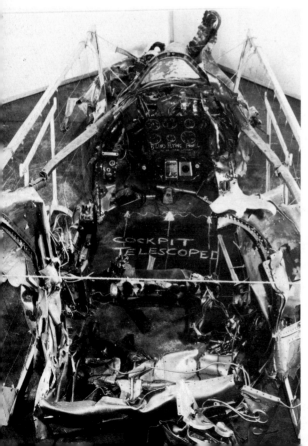

Reconstruction of Whirlwind nose wreckage.

Reconstruction of Whirlwind port wing wreckage (view from inboard and rear).

Reconstruction of starboard wing wreckage of Whirlwind (view from inboard and rear).

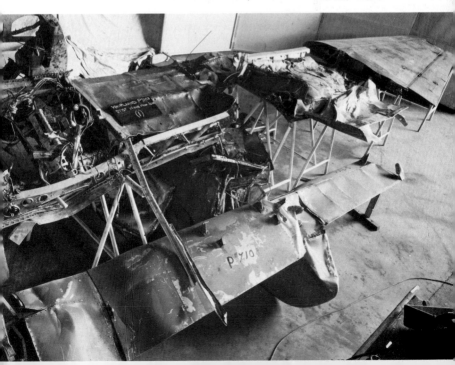

Reconstruction of Lancaster tail unit by trestling and suspension of wreckage, 1943.

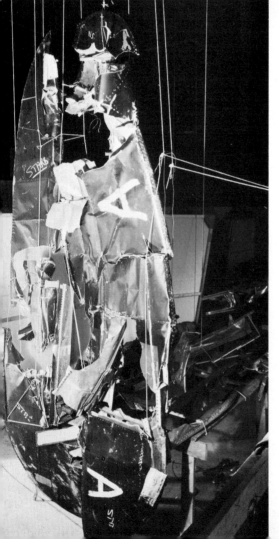

Reconstruction of Lancaster fin and rudder, 1943

# CHAPTER 5

# The Evidence Falls into Place

During 1948, flight research was in progress to determine the efficiency of propellers at high Mach numbers – that is, when the aeroplane is flying near to the speed of sound. Many phenomena occur at very high speeds, and it seemed likely that the development of propeller-driven aeroplanes could be limited simply because of the efficiency of the prime mover at these speeds.

A standard Spitfire Mark 9 was chosen for the test programme, and to assist the scientists in their quest for knowledge, pressure-measuring equipment had been fitted behind the propeller. What looked like two long prongs sticking out on either side of the engine cowling and embracing the diameter of the propeller were in fact an array of forward-facing probes designed to measure the air pressure behind the propeller, and right across the disc.

It was September 1948, and a series of flights was being made at an altitude of 25,000 feet, at varying conditions of aeroplane speed, propeller settings and engine power. The objective was to try to realize a Mach number of 0.84. Already, during eighteen flights, the aeroplane had achieved 0.83.

To achieve the high-speed conditions required for the tests, the Spitfire aeroplane had to be flown into a dive, and it was during the dive, as the aeroplane passed through 25,000 feet altitude, that the test recording would be made. For the test itself, at a given condition of flight, the pilot would press a

65

camera button for the automatic observer panel and, after a second or so, press a second button which activated the test equipment.

Aeroplanes have different aerodynamic characteristics according to speed, altitude, configuration and so on, and often the pilot has to trim or set control tabs to relieve forces on his control column, to make his work easier. And so, for example, instead of, say, holding the control column towards himself to allow the aeroplane to fly level, he could operate a trim so that he could still have the aeroplane flying level but he could take his hands from the controls. In the case of the Spitfire, the scientists desired the aeroplane to pass through 25,000 feet at a certain speed, but trimmed so that the pilot would be able to concentrate on the test. However, the aeroplane had to go from level flight into the dive, and since its speed would then increase, they arranged for the pilot to pre-set the trim tabs for the conditions at 25,000 feet, before he entered the dive. This of course had the effect of putting the aeroplane 'out of trim' before the dive, and also during the recovery after the test; thus the pilot had to hold the control column and push or pull to keep the aeroplane level – nothing particularly excessive, but just an expediency for the test programme.

The details of the test were as follows: climb the aeroplane to 31–32,000 feet, then pre-set the trim tabs for the 25,000 feet condition; as the aeroplane passes down through 27,000 feet, start the camera, then the test equipment; at about 22–23,000 feet start the recovery action; when level, retrim for normal flight, and climb the aeroplane up again for another dive.

The pre-set condition required $2\frac{1}{2}$ divisions of nose heavy trim on the elevator tabs. This posed an interesting situation, because, at the lower speed before entry into the dive, the pilot would be pulling or holding the control column back. To enter the dive, he had to ease the stick forward or reduce his back pressure. In other words, if he were to let go of the column, the aeroplane would suddenly nose downwards. As the aeroplane slowed down and the pilot recovered from the dive after the test, he had to increase his pull, and, of course, again if he were to release the control column, the aeroplane would suddenly tuck its nose down, very smartly.

At 14.12 hours on 14 September the aeroplane took off, but it landed five minutes later with a suspected engine coolant leak. This was dealt with, and at 14.45 hours the aeroplane was airborne again. Nothing further was heard until it was reported that a Spitfire had crashed about eight miles east of the aerodrome. Visitors to the scene found the aeroplane completely wrecked and the pilot dead. His watch had stopped at 15.17 hours, presumably by the shock of the impact in the crash.

There were some witnesses to the closing stages of the flight. It appeared that the aeroplane was seen to perform the first half of an outside loop from a dive. In other words, as it dived downwards, it flew round and under, to be then upside down, flying approximately parallel to the ground. The witnesses then saw the nose drop so that the aeroplane was again in a dive, but upside down. At an altitude of about 5,000 feet, the Spitfire nosed up into a gliding attitude, but still inverted. During this time, a gentle turn to the left was being made. When down to about a hundred feet above the ground, the aeroplane was seen to roll through a right angle, so that its starboard or right wing was pointing downwards towards the ground. Then the witnesses saw a glint of light and movement, as the pilot apparently pushed the cockpit hook back, and this happened just before the machine dropped to the ground.

The weather for the flight was good, light winds and partial cloud in a blue sky with visibility of between twelve and twenty miles.

The pilot had been briefed to make two dives on this flight, each to be at a speed of 370 m.p.h. with engine revolutions at 2,270 per minute. An engine boost pressure of plus 10 lb was to be applied on one dive, and plus 6 lb on the other. Thus the two dives would be made at the same aeroplane and engine speeds, but under different power settings.

The Accidents Investigation Branch, Ministry of Civil Aviation, was informed of the accident but, after a visit by one of its inspectors, decided not to pursue an investigation into the accident. In consequence, I was asked to make an analysis of the wreckage and visited the site on 16 September 1948 and later had the wreckage transferred to the laboratory.

The aeroplane had made a crater in a ballasted path

measuring eight feet diameter by four feet deep, and the wreckage had been strewn to the east of the crater for about thirty yards. One propeller blade was recovered seventy yards to the south of the crater. I searched and scratched around the soil looking for pieces of aeroplane, and, good though the rescue services had been, I was occasionally reminded of the presence of the aeroplane occupant by pieces of human remains, a sight not new to me but one that always made me feel sad, especially if I happened to know the person, as I did in this case.

A narrow trough extended along the ground to the west of the crater, and at its extremity I could see parts of the identification lamp, for the starboard wingtip, the tell-tale green glass. I could see from the nature of the disturbance of the earth along the trough that the starboard wing had struck the ground with the aeroplane diving at an angle of about eighty degrees to the horizontal. The aeroplane had then rotated in the yawing plane onto its port wing, at the same time nosing over in the line of flight. I now knew that the Spitfire, at the instant of impact, was diving towards the south with the starboard wing pointing to the west.

The disintegration of the aeroplane had been so complete that I could not tell at the scene if the machine had been whole when it crashed. This knowledge would have to wait until I had reconstructed the wreckage and made a detailed examination in my laboratory. I could see that the leading edges of the propeller blades had been damaged in a manner showing that the propeller was rotating when it struck the ground, but probably under reduced power. I made the necessary notes, sketches and photographs to ensure that the accident scene had been sufficiently documented for my later reference when I examined the wreckage back at Farnborough.

Just a few days after that sleek and beautiful aeroplane had lifted off into the blue skies, it was entering my laboratory, ignominiously, as a heap of debris, under a tarpaulin on an aeroplane-transporter vehicle.

I spent a few hours laying out the pieces and reconstructing the machine. The engine, meanwhile, had been removed and examined elsewhere, and no evidence was found of any pre-crash damage or defect, neither was there any evidence of overspeeding of the engine. It was back to me and the wreckage.

The entire wing spar, from tip to tip, was laid out. The spar had broken into many pieces, consistent with shockloading during ground impact. The Spitfire wings were attached to the body or centre-section of the aeroplane by a number of substantial steel bolts. I removed these, after marking them so that I could always orientate them with respect to top, bottom, left, right etc. on the aeroplane. There was a row of bolts along the top boom of the spar, and a similar row along the bottom boom of the spar. The upper bolts had been bowed outwards and the lower bolts inwards, telling me that the wing had at some time been loaded downwards. The bolts from both the port and starboard wings told the same story, so the aeroplane had experienced a symmetrical downloading on its wings. The nature and character of the spars' actual disruption at ground impact were quite incompatible with such a loading on the bolts, and of course this told me that the downloading must have occurred during the accident flight. It could not have happened on a previous flight, as the loading would have been too severe not to have affected the pilot, or the consequent damage be missed during subsequent inspections.

Already a picture was beginning to form. Sometime during the accident flight the aeroplane had encountered a severe download situation. Remember those witnesses who described the aeroplane performing that outside loop from the dive; that was the sort of situation which could bring about the bolt damage, if of sufficient magnitude. I knew from experience, and reference to test data, that deformation of those bolts could not be expected unless at least –6g loading had been applied to the wings. I had the bolts checked for material hardness, and they were found to be quite adequate and above specification requirements. In other words, the –6g was real and not masked by some softer bolts, wrongly used, which could have given me a false picture.

Continuing my examination, I identified wingtips, located flaps and ailerons and put the undercarriage legs into position – they were up and locked, as they should be in flight. All detachable panels, for guns, cannons etc., were found. The tail unit was intact and had separated from the fuselage when the

69

latter collapsed at ground impact. Gradually my aeroplane was going together again; soon I could show that it had been complete and intact at the instant of impact.

I moved on to study detail parts now. The pilot was normally held in his seat by shoulder and leg straps, brought together by a common fastening in front of his waistline. All the straps, with eyelets over a common pin in the rear strap, had been damaged in the crash to a degree that left physical impressions of mutual contact between eyelets, showing that the pilot must have been correctly strapped into his seat when the aeroplane had crashed. My examination of the cockpit hood showed that it had not been jettisoned, but from the remains, and the fuselage rails, I could see that the hood had been three-quarters open when the fuselage disintegrated in the crash. So those witnesses did see the hood being opened. The control circuits for ailerons, elevators and rudder were reconstructed, and I could see that all damage and failures had been caused during the disintegration of the aeroplane at ground impact. The same comment applied to the rudder and elevator trim tab circuits.

I now became especially interested in the elevator trim circuit because, of course, it played a special role in the setting-up of the test procedure by the pilot. I was starting to piece together in my mind thoughts on the various items of evidence that could perhaps help me to re-enact the accident flight with respect to the test programme.

The elevator trim jacks could normally be operated over a tab range of two divisions nose up to six divisions nose down. I found the jacks set and jammed at ground impact at $2\frac{1}{2}$ divisions nose down. I carefully checked the circuits and adjacent structure for signs of damage or movement that could have altered the trim tab settings in the crash. I found none. I had to conclude that the $2\frac{1}{2}$ divisions nose down setting was the setting before the aeroplane struck the ground. This was becoming interesting, because the pilot was going to set the trim to $2\frac{1}{2}$ divisions nose down *before* he entered the dive, and retrim after the test. It looked as though something had happened during the test dive, or recovery, and he had not retrimmed. Certainly I could see nothing in the trim circuit itself that could have prevented his doing so.

70

I next examined the oxygen equipment because, with aeroplanes operating at 25,000 feet altitude and upwards, lack of oxygen has occasionally been responsible for pilots losing control and crashing. The examination led me to conclude that the oxygen system had been in order, and turned on, and a check analytically by chemists did not produce any evidence of contamination in the main supply bottle.

Turning now to the automatic observer camera – that is, the one that the pilot would have operated with the camera button – I found that it had been badly damaged. The only means of checking if the camera had been operated seemed to be a visual check of the amount of film on the feed and take-up spools. Such a check, already made by others, suggested that the camera had not been operated, but I did not see the film, and to me the point therefore was unconfirmed. However, I had more positive evidence as to camera state. Remember that, after pressing the camera button, the pilot had then to press the test equipment button. That second button, in fact, fired an electrically operated bomb slip. It could not be triggered mechanically, and furthermore, the electrical circuits were so made that it required the pressing first of the camera button and then of the slip button to complete the circuit and operate the bomb slip. I examined that bomb slip. It had been operated before the aeroplane crashed, so the camera must have been operated too. All of this was saying, also, that the appropriate electrical supplies on the aeroplane were intact and available during the flight.

I felt that I now had enough evidence and information to think about the accident flight and what might have happened.

I had a take-off time, from the control tower, and a probable crashtime, from the pilot's watch. (I had examined it and had no reason to doubt its registered time.) This gave me a total flight time, and it was apparent that the Spitfire had been airborne a sufficient time to have climbed to the test start altitude, dived and continued on down to crash. The elevator trim jack settings showed the pilot had set the trim for the test. The camera and bomb slip buttons had been operated, so the aeroplane must have progressed down the dive through 27,000 feet. That deformation of the wing root bolts had been produced by at least –6g loading, and this suggested a violent bunt or nosing over suddenly of the aeroplane. The most likely time for such a bunt to occur would be either at the commencement of the test dive or in the latter stages of dive recovery. In these conditions, if the control column was suddenly released, the trim of the aeroplane, those $2\frac{1}{2}$ divisions nose down, would take over. The evidence from the wreckage showed that the test dive must have been entered successfully and the dive proceeding satisfactorily. I was left with the conclusion that the bunt must have arisen during the recovery stage.

I had found nothing wrong with the aeroplane, and so the release of the control column could be attributed to some disability of the pilot in controlling the aeroplane. I learned that he had been in good health on the day of the accident flight, and consequently discussed the matter of disability with a friend, the then Wing Commander Bill Stewart, of the RAF Institute of Aviation Medicine, at Farnborough.

Back in 1940, Stewart had been a Flying Officer and one of the small band of doctors making up the RAF Physiological Laboratory at Farnborough, under Dr Bryan Matthews. This small band was, in fact, the nucleus of what was to become the RAF Institute of Aviation Medicine. In the summer of 1941 Bill Stewart went on a test flight in a Boeing 'Flying Fortress'. There had been several accidents to the small number of these four-engined bombers supplied by the Americans to the Royal Air Force, and he was making the flight to study the human aspects of operating the aeroplanes, to see if any explanations for the accidents could be determined from that quarter. The Fortress was flying at about 32,000 feet when it entered extremely turbulent conditions and fell away into a fierce spiral dive. Bill managed to make his way to the rear of the aeroplane, and then its tail broke away with him inside. He escaped by parachute and was the only survivor from the accident.

(Came one of fate's little twists. Keith Swainger went to the scene of the accident to help in the investigation. That left Dr W.D. Douglas empty-handed, and that heap of rocket debris had just arrived at Farnborough. And that was where and why I came into the accident investigation business.)

When I approached Bill Stewart about the Spitfire accident,

72

he had become a Wing Commander and was pursuing a highly successful career, making many valuable contributions to flight safety through his work in the field of aviation medicine. He went on to become first Group Captain and then Air Commodore, and was appointed Commandant of the Institute of Aviation Medicine. Sadly, he died whilst still in the Service. I often wondered whether his adventurous and often dangerous 'physiological' career had led ultimately to his untimely death.

Returning to the Spitfire, we considered that a temporary lapse in efficiency might have been caused by carbon-monoxide contamination, by hydraulic or glycol (engine coolant) fumes or by deficiency of oxygen at altitude.

The Spitfire cockpit, as a type, was particularly free from carbon-monoxide contamination and no accidents had ever been reported from this cause. The fumes from hydraulic fluid, or glycol, were more irritant than toxic, and we had to accept that the pilot would have had ample warning of such trouble and would have been able to take suitable action.

I had not found any evidence of mechanical failure in the oxygen equipment that I had examined, and records showed that the system had been filled before the flight. I had not examined the oxygen mask for fit, to see if any leaks could have occurred. This was, of course, quite impracticable to do after the accident. As far as the subject of oxygen was concerned, the only source of trouble could have been the actual fit of the pilot's face mask on the accident flight itself. This could now only be a matter of supposition, but Stewart had some interesting things to tell me in this respect.

He was thinking in terms of the relative degree of anoxia (lack of oxygen), not a complete lack but a partial starvation, which could easily be obtained if the mechanical system was faulty or a mask leaked quite markedly. The relative degree of anoxia would be similar to that of an individual flying at 18–20,000 feet breathing air, when the actual altitude was 30,000 feet. In our case, the pilot was obviously sufficiently capable of initiating the procedure for the test dive and operating the camera and test buttons on the way down. The relative anoxia theory agrees with this altitude of about 30,000 feet, since it is unlikely that a subject of average resistance to anoxia could reach 30,000 feet and initiate a properly controlled descent without having some oxygen.

Bill Stewart felt that the cause therefore depended on some other factors supervening in order to bring about a complete loss of control. If, for example, the descent had been made from 30,000 feet down to below 20,000 feet in a matter of twenty seconds or so, then if the pilot attempted recovery at this altitude, two physiological variables may have influenced loss of control.

First, in pulling up out of the dive, the pilot would experience some 'g' loading, and this might combine with the after-effects of anoxia. The Wing Commander did not feel, however, that this was a likely answer, because experiments on this aspect had shown that anoxia equivalent to 20,000 feet would cause a reduction in the unconsciousness threshold of about 1–2g that is, a man normally perhaps becoming unconscious in a pull-out involving 6g would have this reduced to approximately 4g with the supervention of anoxia. It would have been highly unlikely that the pilot would have been pulling anything like this level of 'g' loading in the recovery from his dive.

After all this discussion, Wing Commander Stewart then came up with what he felt was a much more reasonable explanation. He felt it might be attributed to what he termed the 'paradoxical reaction to anoxia', akin to overfeeding a starving person.

Tests at the Institute of Aviation Medicine had shown that, if a subject was anoxic, say equivalent to 20,000 feet, when the oxygen supply was restored there would be a fall of blood pressure some three to four seconds later. Some nine seconds later, in the tests, the subject began to make mistakes in a psycho-motor test, and some twenty to thirty seconds after restoration of oxygen the subject actually fainted. If in this case the subject had applied some 'g' loading during the period of paradoxical reaction, he might easily have become unconscious and released the control column. In the case of our Spitfire, because of the particular trim situation, such a column release would have led directly to the negative 'g' loading during the bunting of the aeroplane, and the effects of such loading on an already impaired pilot's circulation is an unknown factor, but since the bent wing root bolts point to a loading as high as –6g,

74

almost certainly the pilot could only be rendered even more incapacitated than before.

Bill Stewart finally concluded that the operative factor would have been the rapid rate of descent (remembering that this was an extremely high-speed dive) since, if the problem had been caused by a mask leak, the rapid descent would cause full reoxygenation of the pilot and would be the equivalent to turning up the oxygen to its full flow level.

Thus our discussions ended with conclusions that the pilot might have been suffering from the paradoxical reaction to anoxia, and the initial anoxic condition might have been only partial, so that he could still perform his duties. During the paradoxical reaction time, the pilot applied some plus 'g' loading during the dive recovery, which caused him to collapse, thereby releasing the control column and bringing about the bunt.

Even without the oxygen deficiency, and/or paradoxical reaction, a bunt of the magnitude necessary to deform those wing root bolts would have severely incapacitated the pilot. Incidentally, the effects of such a bunt could explain the prolonged inverted glide in which no apparent effort was made to right the aeroplane, or for the pilot to abandon the aeroplane. When the witnesses saw the aeroplane roll onto its starboard wing, just before ground impact, and the attempt to slide the cockpit hood back, the pilot was probably making the first attempts to recover the aeroplane to a normal attitude, but there was no height or time left for this to be successful.

The combined efforts of wreckage analysis and aviation medicine experimentation had led to a probable cause being found for the loss of a test pilot and aeroplane. This is a good example of scientists and doctors in the field of aviation medicine pursuing their own specialities and then, when a particular situation arises, being able to augment each other's efforts with their own particular knowledge.

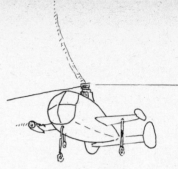

# CHAPTER 6

# *Variety of Problems in 1949*

A wide variety of problems arose in 1949. They included a windscreen failure, which highlighted a design weakness, thereby improving knowledge; the loss of a helicopter due to fatigue in its rotor head, which had resulted from bad engineering practice; the use of escape equipment from a flying wing research aeroplane – the first time such equipment had been used in Britain 'in anger'; a huge container failure underneath a large aeroplane, and a delta research aeroplane with an electrical fault.

On Thursday 3 February 1949 a trainer aeroplane crashed killing two test pilots. The early talk of the cause of the accident was of a cockpit hood and windscreen disintegration, possibly due to a bird strike, followed by loss of control.

An Accidents Branch Inspector paid us a visit at Farnborough on 16 February to give a briefing on the trainer accident, because he needed help. The windscreen and cockpit hood were shattered, regrettably both occupants had been decapitated and the aeroplane had crashed. He had not seen any evidence of bird strike or other external impacting – could we please examine the aeroplane, at present at Wolverhampton, to try to determine the cause of the screen/hood disintegration?

I travelled up to the manufacturers at Wolverhampton on 22 February, had a full day's discussions and studies with the

76

design staff on Wednesday and returned home on Thursday, having arranged for the material to be sent to Farnborough for detailed analysis in our laboratory. Despite the not very encouraging material (heaps of glass and perspex), once some painstaking rebuilding had taken place, surprisingly a story readily emerged.

The windscreens (two large panels, port and starboard) had broken up extensively, and my detailed studies showed that all the disintegration had started on the port side. The nature of the break-up indicated that the port screen had collapsed inwards. There was no evidence of any external impact, and I could only conclude that the failure had been due to air loading. There were no weaknesses apparent in the physical construction of the screen. Calculations and studies of tests that had been made during the initial design stages of the windscreen for the aeroplane showed that estimates of expected loading conditions on the screens during different flight manoeuvres had come out too low. In other words, the screens had been low in strength, and on the accident flight the aeroplane, whilst not exceeding overall flight limitations, had in fact flown into a condition that the windscreens could not withstand. The screens had suddenly burst, and the cockpit had instantaneously 'blown up' by the sudden inrush of air. The cockpit perspex hood had then disintegrated under the high internal loading. The unfortunate pilots, directly behind the screens, were decapitated by the glasswork smashing into them, and a crash was inevitable. As a result of the investigation, stronger screens were devised and the machine became a successful advanced trainer in the Royal Air Force.

The new-styled helicopter, with side wings and a rotor on one side instead of at the tail to compensate for the torque of the main rotor, gave rise to a faster flying machine, so much so that the manufacturer succeeded in capturing the World Speed Record for helicopters in 1948 with a speed of 124 m.p.h. over a three-mile course. Unfortunately the life of the helicopter was to be short. On a low-level run parts were seen to fly off; the helicopter crashed and was totally destroyed, and its crew were killed. A fatigue failure had occurred in the rotor hub. Ironically, one of the crew was a man whose essay entitled 'The

Structural Airworthiness of Helicopters with Particular Reference to Fatigue Failures' had won the first Cierva Memorial Prize.

That fatigue failure had arisen, like so many such failures, not because of strange new materials, clever design or other such things but simply due to lack of down-to-earth engineering practice. Each of the three rotor blades had been held to the rotor head by a large retaining nut, consisting of a shank and a head. The junction between the shank and the head contained an undercut fillet – that is, a recessed groove. The fatigue failure had initiated in this undercut fillet, where machining, not of the highest standard, could also be seen. Examination of the unbroken nuts of the other two blades revealed evidence of fatigue cracking too, in similar locations to the failed nut. It would have been logical and sound engineering practice to have used a generous external fillet, blending the underside of the nut head with the shank. To facilitate this, the top edge of the hole into which the shank was inserted would have then been radiused to accommodate the nut fillet. The likelihood of a failure would have been greatly reduced.

People have devoted a lifetime to studying and researching fatigue and associated matters. I have noticed over the years, having seen many accidents attributable directly to the fatigue failure of an item or component, that the failure was almost always due to the lack of good common-sense engineering practice – machining not as smooth as it should be, holes drilled in the wrong place, and so on. All too often, progress in technology has left behind, in practice, the application of good, sound engineering, which, of course, cannot be ignored where safety is paramount.

On to another subject where this time safety equipment in a very modern-looking aeroplane was involved. It was a tailless machine, the AW 52, popularly known as the 'Flying Wing', because in fact that was virtually what it was, a huge wing with end fins and very little body or fuselage.

Mr Lancaster was a test pilot who was very well qualified, a graduate of the Empire Test Pilots' School. In January 1949 he joined the design firm who had produced the 'Flying Wing'.

Lancaster had flown the 'Wing' on two previous occasions, and on 30 May 1949 took off at 13.53 hours from Bitteswell airfield, near Coventry, to gain further experience in handling the aeroplane. Weather conditions were fair, with moderate cloud; the air was not abnormally turbulent, and the visibility was good. The aeroplane was taken to a cloud-free area over Banbury, and some time was spent between 10,000 and 12,000 feet altitude in quite smooth air conditions.

On reducing height, a longitudinal oscillation set in at about 5,000 feet, at a frequency of about $1\frac{1}{2}$ to two cycles a second. The pilot tried to damp this out by holding the control column firmly. Similar oscillations had been experienced before, and this action had been sufficient to restore normal, steady flight. This time, however, the oscillation continued to build up instead of damping out. The pilot then attempted to ease back the control column, but this had no effect, and the oscillation reached such extreme violence that he was unable to see. He did not know whether this was in fact a blackout, but he came very close to losing consciousness.

The pilot's recollection of the abandonment was vague and incomplete. He remembered thinking of the safety pin in the emergency release handle but could not recall actually withdrawing it. He had a firm impression of the control column being limp and of pushing it forward out of the way. The pilot estimated that some fifteen seconds elapsed between the time of the onset of the oscillation and his departure from the aeroplane. Whatever he recollected, or otherwise, the aeroplane crashed at about 14.30 hours, the cockpit hood was found some three-quarters of a mile from the main wreckage, and the pilot himself left the aeroplane by means of the ejector seat and landed safely. The aeroplane broke up completely when it struck the ground at a shallow angle, and the wreckage was spread over about half a mile of countryside.

The accident was investigated by the Accidents Investigation Branch, and I was asked to examine the wreckage with a particular reference to the abandonment aspect. This was because Lancaster's abandonment was the first use of an ejector seat 'in anger' (that is, from necessity rather than deliberately in a test) over Britain, and the engineers wished to know how it had performed, or what were the faults, if any.

The oscillation problem, a known phenomenon on this type of aeroplane, was handed over to aerodynamicists and flutter specialists to be resolved, because the wreckage examination had given the aeroplane, its engines and all the flying controls a clean bill of health at the time of the accident.

I went up to Coventry on Monday 20 June, to the manufacturer's facilities at Baginton to discuss the accident with the design staff. They then arranged for me to go to Bitteswell where the wreckage had been laid out. Alongside the wreckage was the other specimen of the Flying Wing, so that I was able to read across from my wreckage to a sound aeroplane any details that interested me specifically. I spent until Wednesday examining the wreckage and then after lunch returned by aeroplane to Farnborough. In the few days at Bitteswell I had made an independent check through the wreckage and had also given a lot of thought and attention to the mechanisms concerned with the abandonment of the pilot.

At the time of the accident, ejections were not yet fully automatic as they are today, and pilots were expected to make more than one action before effecting their exit from the aeroplane. In the case of the Wing, the cockpit layout was such that the control column, with its spectacle-type aileron control at the top, moved back and forth between the pilot's knees, the spectacles over his knees, when applications of elevator were being made. To permit any safe ejection then, that control column and spectacles had to be cleared forward. It was to be done in this way: the pilot pulled an emergency release handle, on the cockpit side, after first withdrawing a safety pin (fitted to prevent accidental handle operation); the action of pulling the handle simultaneously jettisoned the cockpit hood and operated a guillotine which severed the flying controls and released a catch; this in turn allowed a spring-loaded device to pull the control column right forward out of the way of the pilot's legs. He was now free to make the ejection, using the face-blind handle of the ejector seat, located above his head.

The idea of the face-blind handle was twofold and novel. It provided protection for the pilot's face as he emerged into the airstream, and it also placed both arms together and upward in front of the pilot to afford further protection and to prevent their flailing about in the airstream. As he pulled the blind downward, a cable would cause a cartridge to be fired, and the seat ejection was then initiated. Once clear of the aeroplane, the pilot had one more action: to separate from his seat. Immediately below the emergency release handle in the cockpit was another handle, provided for jettisoning the cockpit hood only, for such things as maintenance or flight emergencies not requiring the use of the ejector seat.

After the accident, the pilot could not be sure of his actions, whether in fact he had withdrawn the safety pin and so on. My examination was intended to clarify the situation. Having made as detailed a study as possible at Bitteswell, I then brought away parts for closer examination at Farnborough. I had first satisfied myself about the integrity of the flying control circuits in view of the pilot's comments about the column being limp and being able to push the column back before his abandonment. I found no evidence whatever of any pre-crash failure, separation or damage to the circuits. I found the appropriate section of the circuits which embodied the guillotine cutters and saw that the rods had been severed by the cutters. The circuit had been so designed locally as to ensure that short lengths of steel rods would always be in position beneath the cutters for severance if required. I noted that the port rod had been cut outside the rod length, and the starboard rod just forward of its rear end.

I now looked at the emergency release equipment and could see that the handle was still in its housing, and the safety pin had been sheared whilst still in position. The pilot had obviously not made a pin withdrawal at the start of his abandonment actions.

The hood jettison handle was out of its housing, and a close study of the housing and handle components showed that the handle must have been in the withdrawn position when the aeroplane struck the ground. The pilot must have pulled this handle. The action of pulling the emergency release handle would have been to move levers in a box, to operate hood jettison, cable cutter and control column snatch. I examined the levers closely and found various scores and marks to show that the cable-cutter and snatch levers had been moved during impact, after the lever box had been distorted. However, I was satisfied that the hood jettison lever had been moved from

'Safe' to 'Release' before the aeroplane crashed. I was able to conclude that the guillotine had been fired correctly but that the circuits had been located at extreme travel positions when severance took place. The whole could be shown to have occurred during airframe disruption during the crash. The control column snatch unit damage gave me clear indications that the unit had not been fired before the aeroplane crashed. The hood jettison mechanism condition confirmed that the hood had been jettisoned in flight.

My general conclusions were that I had seen no evidence of control circuits failure; I could see no evidence to account for the pilot's impression that the control column was free during his abandonment, and apart from the actual operation of his ejector seat, and the hood jettison handle, the pilot had not touched the emergency release equipment – it had not been operated.

As a result of these examinations, arrangements were put in hand to make the whole ejection process fully automatic, with just one action by the pilot – that of pulling the ejector seat blind handle. Everything would be sequenced to that. Wreckage analysis had played a part in the early days of the development of one of aviation's now most successful safety devices which has saved thousands of aircrew lives.

I was able to keep in touch with this side of the business because every ejection, however successful, was to be fully investigated, for any lessons that might be learned. For example, it was not sufficient for a pilot to eject and arrive on the ground in one piece. There may have been minor snags or hitches in the escape process that did not mar the event but perhaps modified the manner of the final outcome.

Part of the ever-continuing programme of devising and developing means of carrying heavy military equipment by air included the fitment under large aeroplanes of detachable freight containers. These were to be capable of being released in flight and descending by parachute.

One particular container, aptly named a paratechnicon, was so designed that vehicles could be driven into it, the front and rear closed off by huge bowls or fairings, and the whole assembly then fitted to an aeroplane. After the descent by parachute, quick detachment of the front and rear fairings would enable the vehicles to be driven straight into action. The intention was that the occupants would remain in their vehicles throughout the whole operation.

Flight trials were proceeding, and on Monday 26 September 1949 the aeroplane fitted with the paratechnicon flew a handling sortie to determine the effects of this huge container on the directional stability of the aeroplane. It should be appreciated that the container was about six feet high and over thirty feet long, altering the profile of the aeroplane quite considerably. Hence the need for the handling trials.

Approximately one hour after taking off, the aeroplane was seen approaching its base aerodrome at about 4,000 feet. It appeared to be flying straight and level when suddenly the paratechnicon became detached. The aeroplane was then seen to climb immediately. The port wing dropped, and when nearly vertical, the aeroplane dived to port. The witnesses on the ground could see that about half the starboard tailplane and elevator were missing, and they saw the aeroplane perform above five climbing and diving turns to port, before striking high ground about 1½ miles from the aerodrome.

I was asked to make a detailed examination of the wreckage to try to determine why the paratechnicon had become detached from the aeroplane. Investigators had already discovered that it was the paratechnicon that had struck the starboard tailplane and elevator and jammed the controls, leading to the manoeuvres that crashed the aeroplane.

I visited the accident site on Tuesday 27 September, the morning after the accident, and spent two days of examination and discussion with the technical staff of the aerodrome, prior to despatch of parts to Farnborough. The case already emerging for me was to make a detailed study of the huge front fairing or nose bowl, which had completely broken up. I returned to that aerodrome again on Friday to see the wreckage now collected together from the accident site.

By the time the paratechnicon arrived at Farnborough on 18 November I had also received full details of the distribution of the wreckage at the accident site, and this highlighted the fact that the paratechnicon had separated into its major components and all had remained intact with the exception of the huge nose bowl which had disintegrated and had been found distributed all over the wreckage area.

I had already previously determined that the starboard tailplane and elevator had been struck by the paratechnicon skin doors whilst they had still been in position on the main container. The impact had broken the tailplane away downwards, and the elevator detachment had been consequent upon the tailplane failure.

The main container, which looked like a huge metal open-ended and lidless box, had become detached from the aeroplane in a complete state. I could see, from the damage and breakages at its attachment points, that detachment had been due to a movement aft of the container from the aeroplane. The rear bowl had fallen away from the main container following failure of its upper attachments, due to loads on the bowl from forward. A picture was already forming in my mind. Something must have happened to the nose bowl, and air had entered the main container with the rear bowl in position and literally blown the assembly away from the aeroplane. Then the rear bowl had become detached.

I knew now that my interest must centre on that nose bowl. It had disintegrated into about a couple of dozen pieces, and I was readily able to reconstruct the bowl from these. I then, inch by inch, studied all the broken edges of the sheeting that made up the bowl, and prepared a drawing with all the fracture characteristics marked and indicated. That drawing looked something like a map, with arrows tracing out the directions of progression of the various roads – or failures. It was a matter now of 'backtracking' through the failures and local sequences to arrive at the starting-point for the whole disintegration.

The bowl as a unit was held to the main container at four points – namely two upper attachments made of quick-release clips, and two lower attachments made of open hinges, so that a downward rotation of the bowl about the hinges would permit complete separation from the container. I examined all of the attachments and found that the damage to each one was quite distinct in character and different from its fellows, and this told me that the nose bowl had not become detached as a complete unit but had broken up whilst in position at the front

84

of the container. I looked around for any signs that the bowl had been impacted and that this had caused the break-up, but no, there was no such evidence. That bowl had, in my opinion, disrupted under air loads. In fact that detailed examination of all the pieces had presented a very tidy picture. The bowl had separated initially along the longitudinal rivet seams on the bowl centre line, and all other failures had originated from this centre line, in a very symmetrical pattern.

Reference now to the manufacturer's tests on a wind-tunnel model of the paratechnicon showed that, in all normal conditions of flight, suction loads would be present on the sides of the nose bowl. If these loads were sufficiently high, they would cause failure of the longitudinal rivet seam just as I had found. I then learned that the tunnel tests had indicated that the local suction loads on the bowl sides had been underestimated in the original design stage by three to four times.

Came the moment of truth. As a result of my examination, together with the results of the wind-tunnel tests, and strength calculations that had been made after the accident, the conclusion could only be that the initial failure had been the bursting of the nose bowl structure due to the imposition of loads in flight greater than those assumed in the design. The failure could have been expected at an airspeed of about 275 m.p.h., with the aeroplane yawed not more than three degrees – that is, almost straight, not a particularly excessive condition for the aeroplane. All the rest of the detachments and damage followed logically. This was a case of 'back to the drawing board', and indeed strengthening of the nose bowl was called for on other containers before any further flying could take place.

When I returned to Farnborough one evening, I learned that Sammy Esler (a well-known test pilot) had been killed during the day on a flight from Farnborough in one of the new delta research aeroplanes, the Avro 707. Only days earlier, this machine had attracted much attention at the Farnborough Air Show. I was awaiting delivery of the paratechnicon, so next morning, Saturday 1 October, found me in the thickly planted fir plantations to the west of Blackbushe. I spent

Saturday, Sunday and Monday in those trees studying the delta; then it was moved to Farnborough on the Tuesday.

The aeroplane had crashed about ten minutes after taking off on a test flight from Farnborough. Over Blackbushe it descended vertically in a flat spin into the trees. It was readily apparent that the delta had landed in a level attitude whilst spinning to starboard at low forward speed. No trees outside the immediate vicinity of the machine had been damaged, and they were all sufficiently tall to have been struck had the aeroplane made any other approach to the area. There had been only slight ground penetration and, apart from local destruction of wing spars at the centre section by fire, the basic structure of the wings and the fuselage had remained intact.

A search of the countryside did not reveal any portions of the aeroplane, but we did find all components, ultimately, at the immediate accident site. Although the pilot had been removed from the scene before I arrived on Saturday, I could see that the cockpit was shut when the aeroplane had crashed; the pilot's harness had been burned, but the release box was still locked, so Sammy Esler had made no attempt to leave the aeroplane. When we lifted the wreckage for removal to Farnborough, it was held on the crane, clear of the ground, so that I could make a general examination of the underside before it was further disturbed. I could see that the air brakes under the wings had been in the fully extended position when the aeroplane struck the ground. The fuselage air brakes at the rear were about one third open.

Back in the laboratory, we removed the engine and found that it had been rotating at ground impact and appeared to have been operated satisfactorily during the accident flight. The studies at the site and in the laboratory had shown that there was no structural problem, and the engine was in order, so I directed my attention towards the flying control surfaces, their operating circuits and the airbrakes, both underwing and fuselage. This aeroplane was one of the early delta machines and was fitted with both ailerons and elevators; later types had combined surfaces called elevons.

My examination of the elevators and their circuit showed that all the detachments and damage had occurred at ground impact. Similarly I could find no evidence of pre-crash failure, defect or detachment of the ailerons and their circuit. The fin and rudder became detached as a unit as the aeroplane crashed through the trees. I could see no signs of excessive rudder travel before impact. Thus all controls and their circuits were cleared and could not have been responsible for the aeroplane spinning down.

There remained the air brakes to be studied. The controls for the underwing and fuselage brakes consisted of two control levers in a single housing. The levers moved horizontally fore and aft on toothed quadrants, the forward position corresponding to brakes closed and the aft position to brakes open. The lever for the wing brakes was found after the accident at the forward end of its quadrant – that is, where it would be for wing brakes up and closed. The fuselage lever was jammed in the sixth socket of the quadrant from the closed position. I eased the levers out of these positions during my examination and revealed unburnt portions of the quadrant, showing that the levers were in the 'as found' positions before the ground fire. An indicator plate with narrow slots embraced the quadrants and levers, and during the crash it was distorted and broken away. I closely examined the slot edges and found impressions to indicate that the levers had, in fact, been where I found them when the aeroplane crashed.

I now turned my attention to the wing air brakes. The brakes were hinged along their forward edges and were raised and lowered by a pair of hydraulic jacks, one to each brake. When the aeroplane struck the ground, the brakes were knocked away; the hydraulic jacks were fully extended at the time and their piston rods were bent through ninety degrees at the jack body. So the wing brakes were fully down at impact. The fuselage air brakes were intact and undamaged and were about one third open on the fuselage. I could not move them, but when I disconnected the pins between the mechanical links and the hydraulic jacks, the brakes were free to move. This showed that the brakes had been held hydraulically in position. I checked and found that the brake position 'as found' approximated to the lever setting in the cockpit.

From all my examinations there was just one discrepancy on the aeroplane that had been present before the crash: the pilot's selection lever for the wing air brakes was set to 'close', but the brakes themselves were in the 'open' position.

I considered the problem, and found that it could only be resolved in one of three ways: the pilot had selected the brakes down, and the lever had been carried forward by inertia loads in the crash; or the pilot had selected the brakes down in flight and, immediately before the impact, had smartly selected them back up; or finally, the air brakes had moved down in flight without any pilot selection, due to a failure in the brake control circuit.

The first two possibilities were unlikely, as they both required pilot selections, and it was known from previous flights that marked nose-up trim changes had accompanied the lowering of the wing brakes. Trimming out had been necessary, and I found from my examination of the trimmer settings that they were acceptable as evidence and indicated no selection to counter the brake lowering. There were other reasons too why I could not accept these first two explanations, and so there was left just the possibilities that the brakes had lowered following a failure of the control circuit. Experimentation by the manufacturer showed that an electrical failure in the circuit at the pilot's control would, without operation of the pilot's lever, lower the brakes. Unfortunately, despite extensive studies, I was unable to locate any pre-crash fault, largely because the fire in the crash had badly damaged all the relevant components.

We could not then close the case completely but, with the wing brakes as found, particularly if lowered against the pilot's knowledge, the sudden nose-up trim change could certainly have set the scene for a stall and spin, as appeared to have happened.

# CHAPTER 7

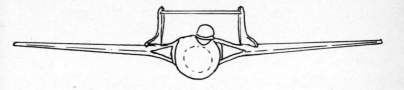

# *Airshow Disaster*

The new aeroplane was progressing well and its flight trials were very satisfactory. It was capable of exceeding the speed of sound in a dive, and in the year 1952 such a phenomenon became the highlight of displays, particularly the SBAC Show at Farnborough.

This new aeroplane, with two engines buried side by side in the central body, behind the crew compartment, had swept-back wings and twin tailbooms supporting a high tailplane between the large fins and rudders. It was the De Havilland 110. Two prototypes were flying in September 1952, WG 240 and WG 236. The former was painted all black, and the latter silver. Both were engaged on the development programme, and WG 240 was including the SBAC Display in its flights.

On Saturday 6 September 1952, however, for technical reasons, WG 240 could not fly, and so WG 236 was scheduled to appear. The pilot was to be John Derry, a De Havilland test pilot, and his observer was Tony Richards. This team had taken the 110s through most of the test flights, and their opener for Farnborough was a high-speed dive, aimed at the airfield, such that they directed sonic bangs at the display area as they went through the speed of sound. The watchers on the ground would be greeted by a double bang, followed by the appearance of the aeroplane.

Saturday the 6th was an ideal day for such a demonstration.

It was warm and sunny with a clear blue sky. At about 3.25 p.m. over 100,000 people were massed together at Farnborough, mainly on the hillside to the south of the aerodrome. Heads were tilted upwards as all awaited the arrival of WG 236. They were not to be disappointed. The double bang preceded the appearance of the silver aeroplane streaking downwards towards them. John Derry levelled the 110 and flew along the runway in front of the crowd. He climbed slightly over the famous Black Sheds and, banking to port, flew a large-radiused turn to the north of Farnborough at about 450 m.p.h. He approached the aerodrome again, over Cove Radio Station, and headed directly towards the masses of people on Cove Hill.

At this time I was in my garden to the north and west of the aerodrome, but inside the aeroplane's circular flight path. My view of the flight was not continuous because of the low altitude of the aeroplane, and trees and buildings between myself and the aeroplane. However, as it flew in over the Cove Radio Station, I had a clear view – I was about $1\frac{1}{4}$ miles away, and viewing from the port rear quarter.

Suddenly the aeroplane reared upwards, disintegrating as it did so. I felt numb because, although remote from the scene, I knew from my witnessing of the flights earlier in the week that the aeroplane pieces must be flying towards that hill and those people massed there. To me the accident was obviously not survivable for the crew – they would have no chance or opportunity to eject, but what of third-party victims?

Later that day I heard the worst. John Derry and Tony Richards had died in the wreckage of the detached cabin of the aeroplane, and twenty-eight onlookers had been struck down and killed by flying wreckage, principally by one of the engines which had flown high over the aerodrome and then plunged down into the people below. Sixty other visitors to Farnborough that day were injured in the accident.

Over 100,000 pairs of eyes witnessed the disintegration of the aeroplane, and so it was concluded that the story of the disaster must emerge readily. In fact, requests were made over the display address system, and later in the media, for witnesses, photographs, films and indeed pieces of wreckage which had disappeared from the scene.

At the time of the disaster I was no longer working on aeroplane wreckages. Some two years earlier the local powers that be had felt that my career and its advancement could best be served by transferring me to other work. That Saturday afternoon in September 1952, as I watched the 110 aeroplane die before my eyes, I wondered, enviously, if that be the right word, who would be making the detailed examination and analysis of the wreckage. On Tuesday, however, I was summoned to the Head of the Division in which I was then working and was told that the Head of the Department, Dr P.B. Walker, had decided that I was to take over the examination of the wreckage of the 110 and that a friend and colleague, who had previously done a stint of accident work, was to help me. The Accidents Section was still in existence, but Dr Walker had decreed that I was to do the job and, in fact, the section itself was not involved in this aspect of the investigation. I later learned that the then Chief Inspector of Accidents, Sir Vernon Brown, had played not a small part in the decision.

On the Wednesday we moved over to a Bessoneau Hangar, a wooden structure, canvas covered, sited on the south side of the aerodrome. The wreckage had already been collected together, and my task was the identification, reconstruction, examination and analysis, with the simple request, 'What broke first, and how?' The why would come later. The aeroplane had broken up into hundreds of pieces, so the task would not be easy.

The first few days, as always, in this sort of task, were taken up with identification and jigsaw puzzling on a three-dimensional scale to recreate our subject. I decided to work on through the first weekend, taking an early opportunity of making some examination quietly. This was a new aeroplane, there was a lot at stake, and any early information would be gratefully received by the designer. I also liked a few hours alone with my wreckage, as the reader will already know.

On that Saturday afternoon I decided to make a study of the main separation of the starboard wing. This in fact came out as a very neat solution. Unknown to me, I had already put my finger on the starting point of the whole disintegration. Probably experience had shown that I could readily count or discount certain evidence, and to the outsider I appeared to move straight in. Certainly on that Saturday afternoon I had not intentionally selected the starboard wing failure, or thought that it might contain the initial failure in the accident.

The days went by and we waded through the wreckage, first producing the basic evidence of failure or damage, then forming local sequences of happenings, and finally producing an overall sequence of failure. Of course the starting-point of this was the first item to fail in the accident. I knew now that the area I had looked at on Saturday afternoon was the vital one, and of course, having already examined it in detail, I was able to say exactly where the whole disintegration had started. We now had reached the why stage.

Each evening, after a day with the wreckage, I read through page after page of witnesses' statements, in the hope that some clue might emerge of value to me. These studies would go on till one or two o'clock in the morning. I recollect looking through at least twelve hundred statements, and hundreds of photographs, all supported by letters, in which witnesses felt certain that they were providing the vital evidence. In the event, when my sequencing was finally completed, it transpired that fewer than a dozen witnesses had told stories that coincided with the now known facts of the disintegration. They all described correctly what they had seen but, by a quirk of circumstance, all those thousands of people saw the accident *only after it had started*, and the few who did get it right were over near to Cove Radio Station, and nearly under, or to the starboard side of, the aeroplane as it approached the aerodrome.

One of the few persons out near the Radio Station was a professional photographer, and he had captured the accident scene on cine-film. He extracted what he thought was the obviously relevant section, had it published in a well-known glossy magazine and made it available to the accident investigators. Examination of it was one of the tasks undertaken by the Accident Section Leader and a German scientist who worked for him. I walked into the Section one day to see what they were doing. A huge sheet of cardboard was surmounted by rows of still prints from the film. There were notes and lines drawn on each frame, from which rates of roll and other information concerning the aeroplane's movements were being deduced. And of course it could all be interrelated on a time-base since the speed of the film was known. There was only one thing wrong – they had not got the full film and in consequence were, unknowingly, analysing in great detail a film of the accident that started far too late in the sequence. I knew already, from my wreckage analysis, that their film started late in my sequence. I told them to obtain the whole film. They did, and promptly began to fill in the information vital to the investigation.

What then had happened to the aeroplane as it flew in over Cove Radio Station? Derry had the aeroplane in a banked turn and would have had to level the wings and raise the nose to fly up over the spectators now ahead of him. To do this, he would have eased back on the control column and also moved it to the right. This latter action would have raised the starboard aileron, and the effect of this would have been to introduce a 'twisting form of loading through the wing structure'. The pulling back of the control column would have introduced some 'g' loading, or upward bending loading, in the wing structure.

It was the combination of these two loading conditions which brought about disaster. Their effects were brought together at the leading edge of the starboard wing, just outboard of the integral fuel tank. My analysis showed that compression buckling had developed around the leading edge, and the associated complementary tension stress across the buckle had caused the skin to split. It was from this separation that the major detachment of the wing developed. The wing was torn off from front to rear, just like tearing a piece of folded paper, starting from the fold. The aeroplane started to respond to this sudden disturbance, which, coupled with the loadings already present, set off a chain of actions which included detachments of port wing, engines, tailplane, cabin and other pieces. The two wing detachments were in such rapid succession that they would have appeared, to a witness, to have been near simultaneous. The result of the loss of the outer portions of swept-back wings would have caused an aerodynamic centre of pressure shift forward, such that the aeroplane then nosed upward. It was the consequent

93

increasing 'g' loading which had then continued the detachments seen by the vast majority of the witnesses. The wings to them appeared to be symmetrical still, so engines, tailplane and cabin featured high in their memories of the disintegration. All of this takes time to describe, but it actually occurred in less than half a second.

I had samples tested of the material from around the origin of failure, at the starboard wing leading edge, and all confirmed that there was no problem of material deficiency. Indeed the very nature of the origin of the failure indicated that we were looking for a particular loading situation that had exploited the local structure of the wing. I have already explained that this had come about from the combination of 'g' loading and aileron application as the pilot tried to level and climb the aeroplane away from the banked turn.

My examinations had continued through engines, controls, systems etc., to ensure that all was fully understood, concerning the disruption of the aeroplane. This phase produced the angle of the starboard aileron for the instant of wing failure. Together with the study of the flight path, angles of bank and so on, we also had the probable speeds and 'g' loadings. Was all of this sufficient?

The firm arranged a structural test at Hatfield and introduced all the features determined during the examination. A non-flying version of the 110 had been fitted into a huge test rig. By means of rods and levers and hydraulic jacks, the wings of the aeroplane could be subjected to the loading conditions experienced in flight.

It was with very mixed feelings that I saw a compression buckle develop within two inches of my own start point in the wreckage. We had reproduced the accident conditions precisely. Contrary to some reports, and authors, I had not marked the test wing with chalk, where I expected failure to occur. There was no need: all concerned knew where the wing had failed in the accident.

I experienced an inner elation of triumph, but equally I was sad, because the work had all shown that if another 110, identical to WG 236, had flown a replica of the Derry Saturday flight, there was no reason why another disaster should not have occurred.

94

The one difference in the local wing construction, between WG 236 and WG 240, which had flown the display during the week at Farnborough, was the absence of an aerodynamic fence on the outside of the wing leading edge of WG 236. This, on WG 240, had been located precisely over the area where the buckling had originated. That aerodynamic fence, intended to deal with air-flow problems over the wing, had obviously provided just the right amount of external stiffening to the leading edge skin to hold the shape and not allow it to buckle. It was likely that during the week WG 240 must have encountered near similar conditions over the Cove Radio Station area, but its fences had been in place. If a buckle had not begun to show, the engineers would have had no warnings of things to come. WG 236 originally had a fence which had been removed as part of the test programme.

As is well known, the De Havilland 110 went on to become the Sea Vixen for the Royal Navy. The loading condition which had proved so critical on this occasion brought forcibly to the attention of the airworthiness authorities the need to include the 'rolling pull out' manoeuvre in their requirements, and for aeroplanes to demonstrate their structural capabilities to meet this requirement. The accident had not been in vain; a lesson had been learned for others to benefit.

For myself, about three months after I had started the examination of the aeroplane, I received a letter to the effect that I could now consider myself back in the Section and that the operative date was the day of the accident.

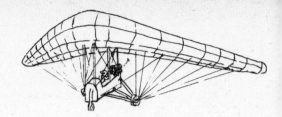

CHAPTER 8

# *Blow up your Subject Every Morning*

From time to time aviation seems to turn the clock back to try to solve a problem, and so it was in 1956. A strange noise could be heard at the Farnborough Aerodrome, either at first light or before dusk, when wind conditions would be at their lightest.

An observer on the aerodrome during these calm periods would have been amazed at what he saw. A large forty-feet-span bluff delta platform wing perched on struts on top of a gondola-like body, with an engine and propeller fixed to the rear, somewhat like an outboard motor on a small water-craft. Flight was being made only a few feet above the ground and, contrary to the usual high speeds associated with delta wings, this aeroplane was flying at less than 60 m.p.h. In fact, people concerned with the aeroplane had no difficulty in following it in motor vehicles.

This was 'Bunty' or, more popularly, 'The Durex Delta', an inflatable machine designed by Marcelle Lobelle of ML Aviation, former Chief Designer of the Fairey Aviation Company and designer of the Swordfish Torpedo Bomber.

It seemed that there was a requirement for an easily assembled, easily transported, quiet, safe aeroplane, operable in every way by one man. Among other things, it could be used clandestinely for lifting agents out of foreign countries. Simply drop your aeroplane, packed in its gondola, by parachute. The

96

waiting person inflates the rubberized fabric wings, fastens a few rigging lines, starts the little engine of about forty horsepower and slowly and relatively quietly putters away, floating over the countryside. At distance the noise might be mistaken for a motor cycle. A good idea, but fraught with problems. And with helicopters coming into their own, this little project was doomed for an early demise.

'Bunty' was the second and improved prototype of the design, the first being 'Loopy'. Both of these aeroplanes had made test flights at Farnborough, and on 26 March 1956, at the end of the day, I was cycling along the perimeter track on the western side of the aerodrome when I heard the putt-putt sound of 'Bunty'. Looking around, I saw it descending to land on the runway. It was immediately surrounded by a group of people obviously associated with the test programme. 'Bunty' then moved forward, took off and climbed away from the runway. The aeroplane flew round to the left, made a low-level circuit and turned to approach for what I thought would be another landing. At about twenty-five feet altitude, whilst on the approach, the starboard wing suddenly dropped and the aeroplane turned and yawed to the right. I heard the engine note rise, as if the pilot had suddenly applied full power, but the descent continued and 'Bunty' struck the runway in a banked and yawed attitude. It then climbed away and, with the starboard wing remaining low, turned through ninety degrees and began descending towards a small copse.

I could see that a crash was inevitable and dashed on my bicycle across the grass to the nearby Police Office at Ively Gate. I rushed in, grabbed the telephone and asked exchange to get the emergency services alerted because 'the inflatable was crashing near C shed'. I then jumped on my bicycle and cycled quickly towards C shed, one of the flight hangars, in time to see 'Bunty' descend into the treetops. I arrived a few minutes later, and there was the gondola with one of my pilot friends sitting in the seat. The wing had deflated and was draped over and around the gondola, a very sad-looking sight. After making a preliminary examination of the aeroplane, although I was not yet officially involved, I cycled away, homewards, pondering the many strange sights I had seen at Farnborough over the years.

Next morning I was given the task of examining the aeroplane for the Board of Inquiry. How could I examine a deflated bag of rubberized fabric, and lengths of nylon and silk cable? Our colleagues at the Research and Development Establishment, Cardington, part of RAE, came to the rescue. Cardington was the home of ballooning and airships, and a section of one of the Farnborough departments was resident there. Dan Perkins, deputy Head of Research and Development, had been concerned with the design, manufacture and even flying of an inflatable aeroplane, and he despatched to Farnborough his chief rigger, complete with the tools of the trade – needle, thread, fabric strip, adhesive etc., and a portable air-compressor.

The rigger made short work of the problem: a few deft stitches, where the branches of the trees had locally torn the rubberized fabric envelope, small patches to reinforce, and we were ready for the compressor. Soon I had an inflated 'Bunty' in the laboratory, ready for my attention. My examination now showed that there had been no structural or mechanical failure in the air. All the rigging lines and control cables were checked and showed no signs of having stretched, or of their knots having slipped.

'Bunty' had been allocated a military serial number, and thus had all the associated paperwork for servicing, maintenance etc. of a service aeroplane. One particular document, called the form 700, is the detailed history of servicing and maintenance of the aeroplane. Everything relating to these aspects is faithfully recorded and signed up by appropriate tradesmen and, where necessary, countersigned by seniors or supervisors. The form 700 also contains a section where the pilot signs to accept the aeroplane as serviceable before flight, and signs after the flight, with notes as to whether it was satisfactory or otherwise on the flight. It is possible then to go through the form 700 and relive the day-to-day engineering activities on the aeroplane, to help keep it airworthy, all of course in strict chronological order. I was able to take the '700' for our 'Bunty' and picture all that had been done, by whom and for what purpose, since its arrival at Farnborough.

Every morning I could switch on the air compressor and

inflate my subject. I could then study and measure and carry out tests and experiments on the control systems, as though it was a rigid aeroplane. I quickly learned that it inflated every day in the same way, such that I could repeat measurements with confidence.

The pilot on the accident flight had been unable to lift the starboard wing after it had dropped during that last flight. One of my tasks would be to determine why this was so. I started with the aeroplane set up in level flight attitude, and measured all control surfaces for all movements made on the control column by the pilot. I then referred to the form 700 and proceeded to work backward through the entries, making adjustments or alterations to the controls or the rigging in reverse order, because, of course, what I had found during my initial examination and measuring had been the ultimate condition of the aeroplane after the work detailed in the 700 had been completed up to the last entry.

The aeroplane was eventually back to the condition for the day it had arrived at Farnborough. By now I had acquired a thorough working knowledge of it and was set to make a detailed study of the controls and the effects for every entry in the form 700 at Farnborough in chronological order. That aeroplane had failed to respond to the pilot's demands and had crashed. I knew nothing had failed or broken but I did know that the controls had been adjusted and altered several times, and wondered if the answer to the accident lay in that quarter.

Before continuing further, a few words about the aeroplane would not be amiss at this juncture. The aeroplane was a simple delta planform wing, with elevon control surfaces fitted at the extremities of the wing-trailing edges. Elevons are control surfaces which can hinge in opposition on either wing, as conventional ailerons, for controlling the banking or wing lifting, or lowering, or together, as conventional elevators, controlling the aeroplane in the pitching, or nose up or down movements, or, of course, in any combination of these modes. Fins were fitted to the top and bottom surfaces of the wing, inboard of the tips. These could be inflated, to be erected or deflated and rolled up and stowed on the wing surfaces; thus various combinations of fin layout could be tried. The elevons were operated by cables from a long control column suspended below the wing. The column was moved by the pilot in the normal convention, and its movement was governed by stops on levers; the corresponding envelope of movement would have been diamond pattern with the diagonals fore and aft, and lateral. In practice, however, the control column fouled the gondola side bracing wires when moved laterally, and unless the pilot released his safety harness, he could not push the control column into the forward portion of the envelope, when operating the column in the fore and aft sense.

The control column operated two large inner levers, which were connected by silk cables to smaller outer levers on pivot posts in the wing at about mid-elevon span. Further cables connected levers at the pivot posts with king posts lashed to the elevons. The inner and outer levers were themselves parallel in the central position of the control column, but not orthogonal, or at right angles to the control cables. Thus, with the column fully aft – that is, elevon up – the cables and levers at the pivot posts approached dead centre.

Later in the life of the aeroplane, the pick-up points for the cables on the outer levers were altered to give shorter arms, or radius of movement, and increased the gearing of the control column, and in consequence the travel of the column in the rear half of the diamond envelope was reduced. The droop of the elevons was altered from time to time by adjusting the lengths of the cables connecting the levers at the pivot posts to the king posts. This meant altering the angle of the elevons without moving the rest of the control circuit, so that whereas previously the elevons were at, say, trail or central position with the control column held at its central position, after drooping, the elevon trailing edges would be lower, with the control column still central. If the pilot pulled the column to centralize the elevons, to have them at their trail or neutral angle, he would find the column had been positioned nearer to him. Here was a means of bringing the column nearer to the pilot but at the expense of altering and reducing some movements of the elevons. This was all to play a significant part in the investigation as it unfolded.

One other area of interest to me was the rigging or bracing of the wing. A pattern of cables attached at various points on the lower surface of the wing was connected to positions on the gondola. By tightening or slackening these cables, the wing could be twisted, when viewed from its tip – that is, increasing or decreasing its angle. In this way, if both wings were dealt with equally, the longitudinal trim of the aeroplane could be altered, or each side could be lifted, or lowered, to trim the aeroplane to fly laterally level.

The object of the flights with 'Bunty' at Farnborough was to develop the aeroplane further, and at the time of the accident the rigging and controls were being adjusted to improve the control ability of the aeroplane. Thus after every flight the pilot would report on the handling capabilities, and the project officer, one of the scientists, would have the ground personnel adjust, tighten or slacken cables, or inflate or deflate fins, and so on, in attempts to cure whatever the pilot felt was wrong.

I was now able to follow through the flight programme and, by reference to the form 700, make the physical adjustments to 'Bunty' in the laboratory, and study the aeroplane for any mechanical benefits or shortcomings that might have resulted.

On the day of the accident, three groups of flights were made. For the first flight, the upper fins were inflated and the lower fins deflated, rolled and stowed. The only other changes from the previous flights were that the outer levers were reduced to $4\frac{1}{2}$ inches, which increased the control column gearing and reduced the effective travel of the column in the rear half of the diamond envelope; and the droop of the starboard elevon was reduced by about seven degrees. This was done to improve the lateral trim of the aeroplane. The pilot then made a number of circuits to left and right to study the effect of the increased control column gearing. He found that the lateral control had deteriorated and he needed greater effort when moving the column laterally. The control column was also too far forward for comfortable flying, and a push force of some ten to fifteen pounds was necessary for level cruising flight. The pilot also noticed that, in flight, the control cables between the inner and outer levers appeared slacker than normal.

Before embarking on the second group of flights the project officer had the following alterations made: deflation, rolling and stowing of the upper fins; a reduction in wing tip angles by

101

one degree to improve longitudinal trim and to reduce that ten to fifteen pounds push on the column; alteration of the elevon droop so as to position the control column further back and, finally, the tightening of the slack control cables. a number of straight flights were made in the new condition, but the wind was too gusty for the pilot to make any real assessment on the changes in control or column forces. He did, however, indicate that the control column was still too far forwards for comfort.

Before the third group of flights was started, the project officer had the control column again brought nearer to the pilot by altering the elevon droop. Several circuits were now made to the left and one to the right. The pilot found that, to keep the aeroplane laterally level, he had to hold the control column one third of its travel to the left. He noticed that the port elevon was vibrating in flight, but he did note an improvement in the longitudinal trim. He now had to push less than two pounds for the level flight cruise condition.

Another flight was now made without alteration to the aeroplane, so that a project officer could watch, from the ground, the control surface movement during manoeuvring flight. The runway was very wide, and the intention was for the aeroplane to be flown along the runway, followed by the project officer on the ground. The pilot found such difficulty correcting the banked turn to the right, on this run, that, when he did succeed, he decided to complete a circuit to the left and then land. He made a normal approach, with a seven-knot wind at about thirty degrees to port off the runway. Now this was when I had seen 'Bunty' as I was cycling along the perimeter track, and, as I have already described, 'Bunty' struck the runway whilst descending starboard wing low – it flew off and crashed into the copse. What I did not know then was the pilot was holding the control column hard to the rear and to the left in his attempts to raise the aeroplane.

Having followed through the flight tests, making adjustments to 'Bunty' as I proceeded. I finished up again with the rigging and control movements as they were when the aeroplane crashed, but now I knew all about that envelope of control column movements and other features. I could now see quite clearly that, whilst the project officer had been making various adjustments to improve the control ability of the aeroplane, those same adjustments had reduced and restricted the available control column movement. On that accident flight the column position for control surfaces central would have been well back in the diamond envelope, and to one side, so that, when the aeroplane dropped its starboard wing and started to descend towards the copse, the pilot had to pull the column to the rear and the left, but there was so little movement available that the aeroplane simply continued downward. How fortunate that the aeroplane was only a slow-flying rubber 'bag', otherwise the consequences of the lack of foresight could have been quite tragic.

# CHAPTER 9

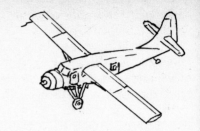

# *A Canadian Interlude*

US Army officers were at Downsview Airport, Toronto, Ontario, in February 1956, undergoing a course of training and conversion on to the U1A, or Otter aeroplane. The US Army were purchasing some of these aeroplanes, and pilots were sent to Downsview, the manufacturer's base, to convert and then ferry the machines on their delivery flights to the United States. The manufacturer was De Havilland of Canada, a famous name and a company famed for its aeroplanes like the Beaver and the Otter, capable of flying into and out of very confined areas and ideal for bush flying in Canada.

On 14 February 1956 three Otters had been allocated for the course, and the flights that day were intended as orientation or familiarization flights. Bill Ferderber, a De Havilland test pilot, was the instructor flying Otter 553252. With him were Major Aaron Atkission, Captain Louis Durand and Captain James Dowling. Flying the Otter was nothing new to Ferderber, who, out of a total of 5,610 flying hours, had piloted the Otter for 480 hours. Major Atkission had totalled eleven hours on Otters, Durand had yet to handle it, and Dowling was to make his first flight in an Otter aeroplane.

Otter 252 took off from Runway 15 at Downsview at 15.21 hours, and at 15.45 hours Air Traffic Control at Downsview received a telephone call that an aeroplane had crashed near Keele Street. All Otters flying were immediately called by the Tower but nothing was heard of 252. At 16.12 hours, a message from the scene of the accident confirmed that it was 252 that had crashed. All the occupants were dead. The flight time for Otter 252 that afternoon had been just seventeen minutes.

Because Otter 252 belonged to the United States, with Army personnel involved, the Americans sent a full Accident Board of Inquiry, including a fair quota of colonels, to Toronto, to carry out an investigation. Technical assistance was provided by the manufacturer. In the course of time, the Board reported that the aeroplane had broken up in the air, and this occurred because the Otter had flown through the wake of a jet aeroplane seen in the area at the time of the accident.

Two months after the accident to Otter 553252, on 10 April 1956, another Otter, this time belonging to the Royal Canadian Air Force and numbered VC-3666, was cleared for an authorized flight from Goose Bay, Labrador. A few days earlier, the port inner trailing edge flap had been damaged and a new flap had been installed. The flight on 10 April was an air test for the installation.

VC-3666 was crewed by three RCAF personnel: pilot, co-pilot and one passenger. The pilot had previously flown 127 hours on Otter aeroplanes and was considered to be very experienced. The weather limits at Goose Bay were in excess of Visual Flight Rules minima with broken cloud at 2,000 feet and overcast at 8,000 feet. Visibility was fifteen miles in light snow, with an occasional snow shower reducing visibility to as low as half a mile.

VC-3666 was delayed after start-up because of such a snow shower and did not receive take-off clearance before visibility improved above VFR minima of three miles. The aeroplane was equipped with skis and took off from Runway 09 at 18.54 hours. The take-off was straightforward, and the Otter continued in a straight climb followed by a slow turn to port onto a north-east heading. The pilot made a normal acknowledgement of the take-off time given to him by the control tower.

Many persons at Goose Bay Airport saw the Otter until it was lost to view in the distance and confirmed that up to that time the aeroplane had been in normal flight.

At 1914 hours a call was received by Goose Bay control tower to the effect that an Otter aeroplane had been seen to plunge to the ground about six miles north of the airport. The witnesses had been the pilot and ground engineer of a private airline Beaver, who had seen the Otter in trouble and proceeded immediately to the crash area.

Two helicopters were despatched to the scene from Goose Bay, and on arrival it was established that all three occupants had sustained fatal injuries at the time of the impact. It transpired that VC-3666 had broken up in the air whilst heading north over flat, moderately wooded countryside. The wreckage fell in deep snow in three main sections. The accident had occurred about two miles downwind of a range of hills 700 feet high. The Beaver pilot reported very moderate turbulence when he arrived over the area immediately after the accident.

On the day following the accident members of the Royal Canadian Mounted Police arrived from Goose Bay and removed the bodies of the three crew members. The next day the wreckage was examined *in situ* by two members of the RCAF Accidents Board and by three members of the Engineering Division of the design firm. After their noting locations of all items, taking photographs and making preliminary examinations of accessible items, the wreckage was collected and returned to Goose Bay RCAF station. Recovery at the site was extremely difficult because of deep snow. In fact a sloping rampway had to be cut down through the snow, and the wreckage, now cut into suitable sections for handling, was hauled on sledges up out of the snow onto the top surface. All personnel were moving about on snowshoes.

About twelve days after the collection of the wreckage, parts of Otter VC-3666 were flown to Downsview, and six days later the remainder of the aeroplane arrived. The major components were reconstructed in a hangar and subjected to a detailed examination. It now became apparent to the combined RCAF investigators and the De Havilland engineers that certain similarities existed between the accident to Otter VC-3666 and the accident to Otter 553252 and that they were too marked to be purely coincidental. De Havilland's therefore launched an extensive investigation into the two accidents, because, whilst the first investigation by the Americans had indicated the presence of a jet aeroplane, and the break-up was associated with the Otter flying through its wake, there were no other aeroplanes flying in the area when VC-3666 crashed at Goose Bay.

One visitor to Canada at this time was R.E. Bishop, Chief Designer at the parent De Havilland Company at Hatfield, in Britain. Mr Bishop had been deeply involved with the Comet Inquiries in 1954 and knew the Royal Aircraft Establishment at Farnborough very well.

Working at Downsview with the Canadian firm was Phillip Halsey, who had moved to Canada from Hatfield. He too had been involved in the Comet story and was no stranger to Farnborough and the Accidents Section.

It was perhaps no surprise then that a letter from Hatfield, with agreement from Downsview, was received at Farnborough, requesting my assistance in the Canadian investigations, particularly in making an independent analysis of the wreckage evidence.

At Downsview on 3 June, Phillip Garret the Managing Director, welcomed me and immediately gave me the freedom of the plant. Anything that I wanted or required was to be available immediately. I then received a general briefing on the accidents from the Chief Designer, Fred Buller, and there and then decided on a general plan of campaign.

Engineering Director, Doug Hunter produced for me a set of handbooks, on the Otter aeroplane, witnesses' statements, photographs and so on. Both accidents had resulted from the aeroplanes breaking up in the air, so I called for copies of the wreckage trails – the maps of the countryside, with the wreckage dispositions marked upon them. From experience I could see that they would provide some useful material, and a little bit of unashamed theatricals, I find, always goes over well, so I described how from these maps we could plot the trajectories of the various pieces of the aeroplanes as they fell to the ground. From this we could ascertain heights and sequences of break-up of the various components. I showed the people gathered round me how this could be done, in a relatively straightforward manner, by weighing and measuring the detached pieces of aeroplane and then calculating the terminal velocities of these pieces – that is, their maximum falling speeds. Then, together with a knowledge of the wind speeds and directions for different altitudes, through which the pieces would have fallen, I showed how to plot out the trajectories.

Two young men from the Engineering Office, Ken Kinsman and Tommy Mann, were brought in and I gave them this briefing and away they went. I guessed how long it would take to produce some results and said that we could discuss them first thing next morning. During the next couple of weeks those two were to provide me with yeoman service during my investigations.

The next day Ken and Tommy were all smiles: yes, it had paid off; they had been able to produce very neat pictures of the accidents, which were in fact remarkably similar, and of course my stock was high already. A pleasing start to the investigation.

I then spent some time touring the plant, to see Otters being manufactured, and was able to see them from the raw material through to the finished product, awaiting first test flight. This was all an extremely interesting experience, even without the investigations in hand. Of course this tour provided me with a clear picture of a 'good' Otter, to aid me in my study of the broken and crumpled remains from the accidents.

The firm had already a fair idea of the sort of thing that had happened to their aeroplanes but, being not normally in the accident business, lacked the confidence to press on without someone with experience to guide them. Although I was making specific studies for them, I also acted as guide and mentor for the whole thing. Whilst I was busy with my examinations and analysis, the firm's engineers were strengthening an aeroplane for some intended test flying, with the possibility of recreating the accident situations, without breaking the aeroplane.

After a few days the wreckage patterns became clear to me. Our structural failures had resulted from sudden and unexpected flap retractions which, together with the tailplane trim settings, had suddenly presented the aeroplanes with complete out-of-trim situations. But before proceeding to the next question – why– a simple description of the Otter aeroplane will help to make clear the story that follows.

The Otter aeroplane is a conventional all-metal, high-wing monoplane, powered by a single air-cooled radial engine. It is fitted with large wing flaps, enabling it to operate into and out of very restricted areas. The wings are strut-braced, and the flaps and ailerons form the trailing edge of each wing. Upward and downward movement of the flaps is obtained by first selecting the appropriate direction on a selector valve and then hand-pumping the flaps to the desired position with a hydraulic system. The flaps cannot move again until hand-pumped, because they are hydraulically locked by a ratchet valve, included in the system for that specific purpose. The ratchet valve basically consists of two spring-loaded ball valves and a central two-way piston. Pressures generated by the hand-pumping are essential to unseat the balls and to operate the piston, thereby allowing fluid flow and flap actuator jack operation. The Otter has a conventional single-fin type tail. Longitudinal trimming of the aeroplane is controlled by a variable incidence tailplane, coupled with an automatic trim tab.

When the flaps are lowered – for example, for take off – if nothing else is done to the aeroplane, the effects of the flaps on the fore and aft balance of the aeroplane would be to make the nose move upward, very markedly. To counter this, the pilot alters the angle of the tailplane, so that its front or leading edge is high, compared with its rear or trailing edge. In other words, if nothing but the tailplane had been set, its effect would be to make the aeroplane nose move downward markedly. If both flaps and tailplanes are set together, their effects cancel and the aeroplane balance remains fore and aft, with no nosing up or down. The aeroplane then feels only the effect in extra lift from the flaps, which is what is required to get out of a restricted area.

At the time of the accidents at Downsview and Goose Bay, a considerable number of Otters had been built; some had flown in excess of 3,000 hours, and the overall total for the entire fleet was approaching 75,000 hours.

My examinations of the wreckage had shown that, at the instant of structural failure in the air, the flaps were up on both aeroplanes, and the tailplanes set fully leading edges up – that is, the flaps and tailplanes were completely in opposition for a balanced trim of the aeroplanes. Either the flaps had been moved upward and the tailplane had not been moved to compensate, or the tailplanes had been moved upward and the flaps not altered.

What had actually happened to the aeroplanes when they broke up in the air was that the wings had broken downward – that is, consistent with the noses of the aeroplanes dropping suddenly. The detaching wings had then driven into the tail units, causing their detachments.

If the flaps were already up, to make the nose drop would require winding the tailplane to its appropriate angle, but this is a slow process and would be almost impossible by the pilot; there was no evidence of a failure of the trimming system that could have suddenly caused the tailplane to change angle to full up. We had to accept that it was the flaps that had moved, not the tailplane. The flaps had to be down to start with, and then be moved upward, suddenly. But normal hand-pumping is a controlled action, and there was no reason whatsoever for a pilot to make such action, without the compensatory trimming of the tailplane. The clue lay in that ratchet valve. If it could be faulted so that the spring-loaded balls lifted off their seats, the 'hydraulic locking' of the flaps would cease and the flaps, if down, would move upward under the pressure of the air as the aeroplane flew along, or if up, and the aeroplane was on the ground, the flaps would simply move downwards under their own weight. However, none of these actions would occur until a selection had been made, on the selector valve.

The story could now read like this. Have a defective ratchet valve and have the flaps already selected and pumped down and the tailplane trimmed appropriately. Simply make an up selection of flap and, because the ratchet valve is defective with a ball not firmly on its seat, the flaps blow up without any hand-pumping. Imagine this happening as the aeroplane is flying along, and the sudden 'out of trim' produced by flaps moving upward and tailplane staying fixed would cause the aeroplane nose to drop, or the aeroplane would bunt, and the resulting download on the wings brings about their failure.

With this possibility before us, excitement ran high and work proceeded along several lines. We had heard that there was available a suspect ratchet valve from a serviceable aeroplane, where the flaps had moved down under their own weight when selected during ground servicing. This was the

110

opposite sense to what we were thinking for the accident cases, but the same principles applied.

This suspect valve was now incorporated on a hydraulic rig in the test department and the system pumped up to pressure. Immediately the dial gauges on the rig showed that the valve was not functioning correctly. An X-ray showed that a ball was off-seated, and a careful strip examination revealed the cause – a small piece of metallic swarf between the ball and its seat. Another ratchet valve, straight from store's shelf, was next put onto the rig. Fred Buller himself had operated the rig to disclose the faulty valve, so this time I was given the privilege of 'playing with the rig'. I had not been manipulating pumps and valves for very long when again the dial gauges showed that my ratchet valve was in trouble. Another X-ray and careful strip, and again small particles of metal were seen under a ball. It appeared that the metal particles were produced when a union nut was screwed up to make a pipe joint during assembly and small parings of metal became detached from the threading. We had found our probable culprit, but would the wreckages support our thoughts for the accident cases? In one case, although we did not find the ball valve actually held open by swarf, we did find samples of swarf in positions to make it a very viable proposition. In the other case, swarf was found in other parts of the hydraulic circuit.

We had all the ingredients for a flap-system failure, and we also had the appropriate consequences in the form of the nature of the break-up of the aeroplanes, and of course the flap/tailplane 'out of trim' conditions. However, as we really wanted to clinch the matter much more firmly, the firm set about making a number of tests on the ground, preparatory to a full-scale air test, with a stimulated flap-system failure built in.

First of all, a simple bypass arrangement was built around a ratchet valve. This incorporated a valve that could be 'cracked open' by different amounts, to simulate different sizes of swarf obstruction under the ball of the ratchet valve. Ground tests were made on a flap/wing assembly to establish 'blow-up' times of a flap with different 'swarf sizes'. Meanwhile an Otter aeroplane was being specially strengthened to be able to fly safely to a condition – far beyond that which it was known had broken the two accident aeroplanes. The firm had taken another Otter and tested it under download conditions in the laboratory to failure, and so we had a measure of the loading we could expect on the test flights.

Gradually all the information was drawn together, and the strengthened and instrumented Otter was prepared for the flight tests. The programme was planned for a series of tests at increasing aeroplane speeds and increasing 'swarf sizes', thereby speeding up the flap up movement time and consequential bunt or nosing down action.

George Neal, the chief test pilot, flew the Otter with Bob Fowler, the assistant chief test pilot, as co-pilot and operator of the bypass system, to impart as near as possible the surprise element of the accident situation. Bob also acted as safety pilot, in case George needed help during a test.

A chase plane, a Beaver, escorted the Otter on every flight, taking air-to-air photographs of the Otter's movements during the tests. A three-way radio link was established between Otter, Beaver and ground control. Everything possible was done to set up a representative and near realistic flight case, safely and fully recorded, so that we could approach, pass through and, if needs be, go beyond the accident condition.

On the ground I joined a full team of engineers, aerodynamicists and pilots, all available to monitor, discuss and advise as the flight test actually proceeded. The Chief Aerodynamicist, Jack Uffen, controlled the whole affair, and after each test run a plot was made to determine how safe it was to continue. This was all coloured by the comments of George and Bob on their inter-com sets, also relayed to us down below.

As the testing proceeded, some of these comments suggested that rather 'hairy' situations were being experienced up aloft. In fact, because of the special strengthening, the Otter did fly through the accident condition. Jack Uffen finally decided to stop the tests – it was now really getting very tricky, and not advisable for George to press on any further.

All our efforts then culminated in success. The introduction of a filter into the flap hydraulic system, and of an interconnection between flaps and tailplane to ensure a state of 'in trim' whatever the flap angle, meant that the Otter

Aviation's first jet fatality: Gloster F9/40, DG204, January 1944. Site of main wreckage.

F9/40, tail unit detached in the air and landed on the roof of the RAE foundry.

Accident site of Stirling LK207, 1944.

First large-scale fuselage reconstruction ever attempted in this country –
Stirling LK207.

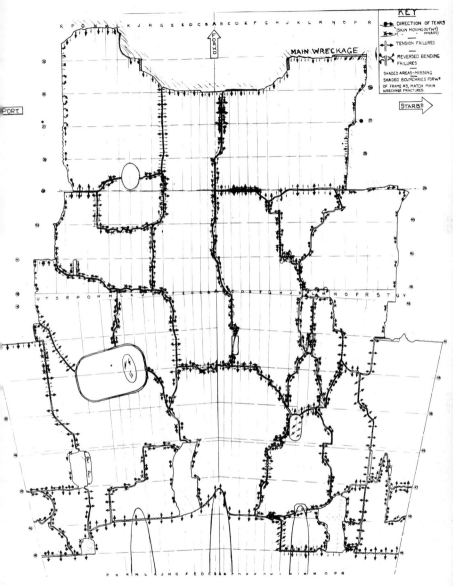

Development drawing of large fuselage (Stirling LK207) with characteristic of breakages indicated.

Reconstruction of V2 from Sweden 1944. Mounting and end frame for engine and accessories.

Venturi from V2 from Sweden.

Reconstruction of warhead of V2 from Sweden.

Steam turbine and fuel pumps from V2.

Paratechnicon under fuselage of aeroplane.

DH110 breaks up
over Farnborough
aerodrome during
1952 Air Show.

Forward fuselage
with crew
compartment strikes
ground in front of
spectators. (*Inset*)
DH110 minutes
before the accident.

Starboard wing of DH110. Arrow points to buckle in leading edge where disintegration of the whole aeroplane was initiated.

Close-up of reconstructed leading edge of the starboard wing. Arrow points to origin of wing failure. Note: author's pencil markings of the fracture progression directions can be seen leading away from the origin.

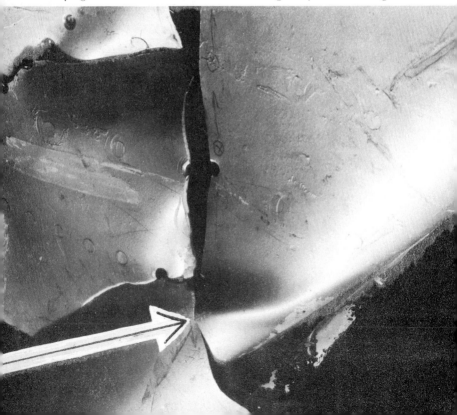

aeroplane, already an excellent machine, would now go on from strength to strength.

And so ended a brief but extremely interesting and well-packed interlude in Canada, and of course, we did have the pleasure of correcting our American cousins. Apparently they had decided that the evidence concerning the tailplane angle in the crash at Downsview was not reliable and in consequence had missed the vital feature of the aeroplane being 'out of trim'.

# CHAPTER 10

# The Americans Cry Help

On 6 April 1958 Capitol Airlines' Viscount N 7437, with forty-seven people on board, left Flint a few minutes after eleven o'clock at night on a short routine flight to Tri-City Airport, Saginaw, Michigan. The flight was scheduled for a duration of about fifteen minutes. The weather at Saginaw was dark, very gusty, with light snow showers and freezing conditions. Witnesses saw the passage of the aeroplane across the sky only as a path traced by cabin lights and the occasional silhouette of the aeroplane itself. The Viscount appeared to have joined the airport circuit without incident but, whilst making a steep turn onto the final approach path to the runway, at an altitude of a few hundred feet, it suddenly plunged to the ground. The aeroplane broke up on impact and was severely damaged in the ensuing ground fire which persisted for an appreciable time. All the forty-seven occupants were killed.

The Bureau of Safety investigators made an initial investigation at the scene of the accident and then reconstructed and examined the wreckage in a large building in Washington, on the banks of the Potomac River.

Early in July 1958 I received a telephone call one afternoon from Morien Morgan, Deputy Director at Farnborough. The Americans had a problem, they wanted some assistance with the examination of the Viscount wreckage, and Morien wanted me to provide that assistance. I flew out to the United States on

Sunday 20 July 1958 and arrived in Washington on Monday morning. In the afternoon I met up with the Americans at the Bureau of Safety, located in a building within sight of the White House.

Before proceeding further, I must fill in the background, to show where my efforts were to fit into the overall pattern of the American investigation.

In 1958 the Bureau of Safety was the official body in America charged with the investigation of civil aeroplane accidents. It was divided into several divisions, each dealing with a particular aspect or phase of an investigation. In general, this was what happened when an accident occurred.

The Investigation Division supplied personnel who made the initial fact-finding examinations and enquiries. It was assisted, as required, by the Technical Division, which provided a general technical service to the Bureau and made available specialists for investigations. The facts and data thus obtained were then submitted as exhibits at a public court hearing, presided over by a member of the Hearing and Reports Division of the Bureau of Safety. This procedure allowed the facts to be 'aired' in public and additional evidence possibly obtained as a result. The whole case was then referred back to the Investigation and Technical Divisions personnel concerned. Analyses were made of the facts, and analysis reports were then submitted to the Hearing Officer, who compiled the accident report. This document, after the customary vetting and discussion, and when to the satisfaction of the Bureau, was then sent to the Civil Aeronautics Board itself. The Board, consisting of five members, all appointed for short terms by the US President, finally gave approval to the report, which was then published over the Board's signatures.

The Viscount investigation had reached the stage of analysis, after the public hearing, when I visited Washington, and it became apparent that differences of opinion and doubts existed as to the attitude of the aeroplane when striking the ground. This indecision also left obscure the nature of the closing stages of the accident flight and hence the possible cause of the accident.

The strongest opinion amongst the personnel concerned, *largely from the Technical Division*, seemed to favour the aeroplane

115

striking the ground at an angle of up to about forty-five degrees and then pitching over onto its back. The minority opinion, *from the Investigation Division*, was that the aeroplane struck the ground nearer to ninety degrees before pitching over. This was the background to the investigation at the time I was asked to examine the wreckage, and it was very obvious from the start that I had been brought in as a sort of arbitrator. Thus, although I would be making my own unbiased examinations and judgements, I would be doing so in an atmosphere of tension because both parties were very adamant and intense and voluble about their opinions. I was, however, given complete freedom of movement and access, and no one attempted to persuade my thinking in any direction, apart from the wreckage itself, of course. I was able to study not only the wreckage but all exhibit material that had been submitted to the Court of Hearing.

Of course the provision of the exhibits was a necessity, as I was having to resolve a problem at a time and place remote from the accident, and much of the wreckage had been dismantled during the Board's investigation.

First I established the salient features concerning pertinent wreckage items; then, with the emergence of a logical sequence of events, I consolidated the facts by further detailed analysis. It became readily apparent to me that a logical damage sequence existed which could indicate the true nature of the impact of the aeroplane with the ground. But now I was faced with a major problem of presenting the facts to the Americans. My sequence *was contrary to that already assumed by either party*, although it more nearly favoured the minority opinion. In other words, I had been brought in effectively to arbitrate between two schools of thought, and I had promptly produced a third story.

The members of the Investigation Division readily accepted my sequences, because they were engineers who dealt with wreckage all the time and could appreciate the finer points that I had produced, but, of course, I had to convince everyone – or, rather, let the people concerned convince themselves.

The plan to do this came about quite naturally. I was dealing with proven facts which could be assembled into a logical sequence of events. All I had to do was to show the

Bureau the proven facts in order of sequence and get their acceptance of them piecemeal, letting them then add up the facts for themselves. The Bureau members were led inevitably through the sequence, without any persuasion by me. There was one leading dissenter – the Chief of the Technical Division. He had made no hesitation in telling all earlier that he and his colleagues were right. My *pièce de résistance* – and, incidentally, the *coup de grâce* for our friend – came with the last area of wreckage to be considered. Of course, by leading him through the previous items and letting him accept what were indisputable facts to everyone, I was setting him up, because the last area of wreckage simply completed the sequence and pointed exactly to how the aeroplane had struck the ground. The Chief walked over to this wreckage, took one look, turned and walked out of the building, very red-faced, without saying a word. He realized what had happened; everyone present clapped and applauded; I had no need to enquire about the acceptance of my findings.

I carried on with some detailed work necessary to complete my notes for my report to be written back in Britain and spent some time going over the finer points of the story with the Bureau personnel, who by now were eager to learn more. The Chief came round next morning to admit defeat. In fact, we became very good friends, meeting up on mutual problems in distant parts of the world on more than one occasion.

What was the story that I had been able to obtain from that wreckage?

A very consistent pattern of damage presented itself to me right across the aeroplane from wingtip to wingtip. The leading edges of both outer wings had been crushed downwards round the wing structure and, in addition, the starboard outer wing had broken off by bending tip downwards about nine feet from the tip. The heavy structure which constituted the fuselage nose flooring forward of the wing had collapsed downwards and rearwards. All propellers had sustained remarkably similar damage. Each engine drove a four-bladed propeller and, when viewed from the front, corresponding blades had been similarly damaged; thus one blade was bent backward along the top of the engine, the side blades had bent downwards and backwards, and the bottom blade was still

117

reasonably straight. The propeller shafts were all bent downwards.

Photographs of the accident site showed that all four engines were in their correct positions relative to the aeroplane. All were buried to a depth of about five feet, having entered the ground inverted, and from the north-east, at an angle of about forty-five degrees. There was no evidence of any engine movement after the initial penetration of the ground. The angle of entry of the engines must also have been the angle of impact of the aeroplane, because the failures of all engine-mounting feet attachment studs showed that the wings and mountings, and hence the aeroplane, had all continued to move forward into the engines, as the latter were arrested. Propeller blade damage also suggested little or no rotation after initial impact with the ground. However, torsion failures of all engine shafts indicated that the engines were rotating, and pointed to the propellers being suddenly stopped by the penetration into the ground.

There was no doubt in my mind that the aeroplane had struck the ground when already inverted. There had been no pitching over, as the Americans had concluded; indeed, the wing-leading edge damage was completely wrong for this. One other area of interest was that the cockpit and front upper portion of the forward fuselage had been detached and was found in pieces to one side of the accident site. The Americans had had much discussion and controversy over this and had finally concluded that the pieces must have been thrown there during ground impact. I examined the pieces and was able to build a picture of the manner of detachment from the rest of the nose section. I arrived at a very different story: the aeroplane, inverted, was drifting to starboard, and the upper nose and cockpit pieces struck the ground first *and were left behind* as the main aeroplane moved over and downwards into the ground.

The more detail I studied, the more evidence there was to show that the aeroplane had struck in the manner I described. The Americans had seen all this information for themselves – I know, because I read their submissions to the Court of Hearing. All that was wrong was the lack of interpretation in detail.

118

My overall examinations had led to the conclusion that there had been no structural failure or separation of any component in flight, that there was no evidence of pre-crash defect or failure of any flying control circuit, and that the undercarriage was down and in position for landing, as were the wing flaps. My task was done. Once again wreckage analysis had shown the way towards the correct solution. The investigators and manufacturers were able to pick up from where I left off and feed the impact attitude of the aeroplane back into the other information they had concerning the closing stages. From this they were able eventually to reach a probable cause for the accident.

I departed for London just ten days after arriving to commence my investigation.

# CHAPTER 11

# The Sea Gives up an Aeroplane

On Friday 21 August 1959 I received a telephone call from the AIB. It appeared that a Victor bomber had suddenly disappeared from the radar screens whilst over the Irish Sea, off the Welsh coast. No report had been received of a crash, but the aeroplane was long overdue. Given the last plot for the aeroplane, its general heading, approximate height and speed, could I make some estimates of areas on the ground or sea in which searches might be initiated? Quite a tall order, although not the first time I had received such a request.

This time, the aeroplane was last known to be at about 50,000 feet altitude and travelling at high speed towards the south-west, but of course at the time of the 'disappearance' we could not know whether the aeroplane was whole or in pieces. I did some rapid calculations, hedging my bets both ways, and telephoned back some general locations that logically might contain the bomber.

During the days that followed, the story emerged. That machine had been a prototype for the latest version of the Victor. It had been lost whilst on a test flight from the Aeroplane and Armament Experimental Establishment in Wiltshire. I also learned that a small coasting vessel, the *Aqueity*, had reported seeing a splash in the sea some eight miles distant at about the time the radar plots of the aeroplane had stopped.

On Thursday 27 August 1959 I attended a meeting at the

Admiralty in London, with my Divisional Head. It had been arranged to discuss the search, salvage and recovery, if possible, of the bomber. Two days earlier small pieces of fibre-glass material had been picked up on the Welsh coast shoreline in St Bride's Bay; we had been able to identify them as part of the bomber's radome.

It was evident now that the aeroplane was out there in the Irish Sea. The meeting discussed the merits of the evidence available – the radar plot, the splash and the radome material, and then decided on a likely area for the initial search. The Royal Navy had available frigates, a special salvage ship and some civilian fishing trawlers. We had been warned of the difficulties of the task ahead. The area was the bleakest and roughest patch of sea around the British Isles, and the water was about 400 feet deep. Fortunately, the sea-bed was described as sandy with a fairly level surface, although tidal movements and currents were to prove troublesome.

The general plan was for the radar/echo surveys to be made by the frigates, and any possible/probable echoes would be investigated by the trawlers. Small items would be retrieved by nets, larger items by the salvage vessel. Any recoveries would be transferred ashore and then transported by vehicles to my laboratory at Farnborough. At the time of the accident I had a small RAF team working with me at Farnborough, and I planned to put one member on the salvage vessel and another on shore to monitor reception and loading for onward transmission of the material from the sea. The Navy, fortunately, had a shore establishment near Haverfordwest, and the Commander here controlled the whole task force.

At Farnborough Dr P.B. Walker, Head of Structures Department, had been tasked with the overall investigation into the accident, and I was responsible for all aspects of the wreckage – its identification, examination and analysis.

By 31 August more radome pieces had been found by the shore search-parties, and I was able to identify these positively as coming from the front radome, the rear radome and amidships, so all of the aeroplane was out there in the bay. I then had an idea – put some samples of radome material, suitable identified, into the sea, in the general area designated for the search. Choose a time when sea and weather conditions were similar to that at the time of the accident, and wait and see where they floated and drifted. If they arrived along the shoreline where we were finding the aeroplane radome pieces, we would then have a notion that the search area planned was right.

On 4 September, during the morning, thirty specimens were put overboard by a trawler, at three locations in the search area. We had used pure aluminium labels riveted to the specimens so that identification could be preserved in the salt environment. We then had to wait. Then the specimens began to arrive back at Farnborough, from Irish police stations and coastguards. The whole collection had gone to Ireland. We made a second attempt, but we never had sight or sound of these specimens, so it is reasonably certain they all drifted into the open Atlantic.

On 16 September 1959 we had received thirty-four pieces of radome, one piece of interior padding from the cockpit, and three pieces of aircrew helmet padding, all as flotsam. So far nothing positive had emerged directly from the search at sea, but not for the want of trying.

I was kept busy receiving, examining or having analysed samples of oil slicks, pieces of debris from the nets, and trawl ropes that had snagged items in the water and had been damaged. It was hoped that I might pick up the minutest vestige of evidence, positive enough to show that the trawlers were in the right place. One day I received a piece of aircrew helmet, and a couple of days later, on Sunday 14 September, I received a cryptic and guarded phone call to tell me that 'they had the owner of the thing I had received a couple of days ago' – a guarded message over the GPO lines to indicate that a crew member had been recovered.

Late though it was, I contacted my friend David Fryer, a doctor at the Institute of Aviation Medicine, who was standing by for such news. I also alerted Commander Flying at Farnborough for an aeroplane to be made available at first light on Monday morning to take David to Wales. I went to bed assured that all was in hand. Then the telephone rang – it was David: he had done some telephoning himself and learned that the sea recovery had been a civilian and, furthermore, a newspaper in a pocket was dated after the day of the accident.

On another occasion I received a box of heavily wrapped bones, reasonably fresh, fished out by one of the trawlers; could they be from a crew member of the aeroplane? I passed them directly to another doctor friend, a pathologist, Peter Stevens, in the Royal Air Force. His detailed report made interesting reading. He was able to tell me that the specimens consisted of half a pelvis, and halves of three vertebrae, all connected by muscle and tissue, which had undergone considerable 'adipocere formation'. But it was not a human pelvis, but was most probably that of a two-year-old steer, butchered and sawn through the middle.

The days went by with highs and lows of anticipation, right through the winter of 1959 until, on 5 January 1960, there came the break we had all been hoping for – but not from the search itself. A trawler, aware of the search, passing through the area pulled its nets and found a small piece of corrugated metal. It was hurriedly despatched to Farnborough, and I was able to identify it as our first piece of bomber from the sea.

Having in mind the trawl line on which it was found, a reconsideration of the search area was now made, and the Royal Navy re-established a datum, in fact near to where the Master of the *Aqueity* had said he had seen the splash. The Navy had not accepted his report completely earlier. It was thought that he could have over- or underestimated distance to splash, and height and size of splash as well – after all, he was not a trained naval gunnery man, so how could he be correct? Had we been waiting all these months because the Navy had decided they knew better than a witness where to search?

On 14 March 1960 the search trawlers found our second piece, the next day came fifteen pieces, and so we had started. Quickly it was established where the main area of wreckage was, on the sea-bed, and a systematic search and recovery programme was then put in hand, with the trawlers queueing and passing through the area, using Decca navigation to ensure complete and accurate coverage of the sea-bed.

From now on, recoveries began to flow steadily, and they continued at regular intervals through to 22 November 1960. To speed deliveries to me at Farnborough, at the start of the recoveries, I arranged with Commander Flying at Farnborough for airlifts from RNAS Brawdy to Farnborough. In the event, forty-six flights were made to Farnborough before the task was completed, and we used our Canberras, Varsities, Hastings, Beverley, Bristol Freighter, Shackletons and Devons. I also had a large number of very heavy duty canvas bags or sacks made up that would contain wreckage in convenient quantities for handling, and these were sent out to the trawlers. We began to receive wreckage within hours of recovery from the sea and still very wet.

The weather was very hot and sunny, and after every delivery the laboratory smelt like a fish market. But there were compensations. Occasionally a wooden box was received labelled 'Personal' to me. The trawlermen were sending me a present from the sea – very large cod for the investigators to eat. Those trawlermen did yeoman service. Up until the time in March when the wreckage was discovered, they had shot and hauled their nets 1,367 times, often in foul and adverse weather. But after the discovery, the fleet of trawlers, which varied between eight and sixteen ships, with a salvage ship providing a back-up, carried out 11,069 hauls. Throughout the whole operation some forty ships, naval and civilian, and about fifteen hundred men were engaged on the search and recovery of the bomber pieces.

As each trawler hauled its net, the crew radioed ashore, to a team of Wren radio operators, the number of pieces of aeroplane retrieved. Thus a tally was made of the pieces recovered and sent to Farnborough. When the last net was shot and hauled, and its contents counted and despatched to me, we had amassed 592,610 pieces of wreckage on the laboratory floor. Individual pieces ranged from a few ounces up to pieces of engine weighing five hundredweight.

As every bag was offloaded at Farnborough, each piece of wreckage had to be handled, identified and examined, because I was attempting to reconstruct the aeroplane. This was inevitably a very slow process because I was dependent on the rate of recovery from the sea, and on the obvious random nature of that recovery. At the end, we had received about seventy per cent, by weight, of the aeroplane, and had been able to rebuild enough of the many components to present a fair representation of the aeroplane. To illustrate the completeness of the gleaning of the sea-bed by the trawlermen,

124

I received the co-pilot's wristwatch, without straps, and was able to establish a reliable time of impact of the aeroplane with the sea.

Gradually a picture of the aeroplane's condition at the time of impacting the sea emerged. I was able to show that it had struck at a high forward speed whilst diving sharply, the right way up. It had been structurally complete and intact, apart from the wingtips and bomb doors which, it could be shown, had become detached late in the descent. The aeroplane had been in a clean configuration – that is, undercarriage, flaps and dive breaks were all retracted, but there was one notable exception: the nose flaps along the wing-leading edges had all been extended. I could also show that fire had not occurred at any time during the accident flight and that electrical power had been available right up to the instant that the aeroplane had struck the sea. All the engines had been rotating at high speed and all throttle levers had been found set at high power settings.

There was evidence to show that oxygen had been available to the crew. Both the pilot's escape hatches had been released before the aeroplane hit the sea, and the pilot's ejection seat had left the aeroplane in the correct manner before sea impact. All automatic mechanisms had functioned, and the pilot had separated from his seat. The co-pilot's seat had not left the aeroplane, and the evidence from our examination indicated that the seat was still occupied but in process of ejection. The rear crew seats (non-ejecting type) were barely recognizable, but sufficient strap material was found to indicate that all three crew members were probably still sitting in their seats when the aeroplane crashed.

The bomber had descended at high speed from very high altitude, partially breaking up during the latter stages. The only feature of note in its otherwise clean configuration was the extension of the wing nose flaps.

The wingtips' wreckage was assembled and examined in detail, and the most significant feature disclosed was the detachment of the pitot tube from the starboard tip, before the tip had disrupted in flight.

One other feature from my examination now also became worthy of consideration. It was the condition of the auto-mach

125

trim actuator. The longitudinal trim of the aeroplane at high Mach number is maintained automatically by an item called the auto-mach trim actuator. This actuator is electrically operated and fitted to the elevator control circuit so that it causes the elevator to move upward as it extends, and conversely to lower with actuator retraction. This is independent of any demands by the pilot on the controls. When the aeroplane had smashed into the sea, that actuator had jammed at a setting equivalent to a Mach number of 0.855 – that is, the actuator was virtually fully retracted. Being an electrically operated irreversible mechanism, this setting could be accepted as that present prior to the sea impact. This particular actuator, I learned, had been calibrated for operation from 0.85 to 0.97M.

The features that appeared of most significance from the examination of those half-million pieces of wreckage were as follows:

1) The aeroplane had descended into the sea at very high speed.
2) At the time its nose flaps were extended.
3) At the time its auto-mach trimmer actuator was almost fully retracted.
4) The wingtips had disrupted late in the descent.
5) Before the starboard tip disrupted, its pitot tube had been detached.

Had we got the ingredients for disaster in this evidence? What would happen if the pitot tube became detached? What would be its effect? Why were the nose flaps extended – these are definitely not a normal feature of high-speed flight – and why was the auto-mach trim at the low speed end of its range?

A few words about the normal functions of all of these items and all will suddenly become clear.

A pitot tube is a simple device usually fitted to an aeroplane well clear of interference such as a wingtip, because its task is to monitor air pressure, which is then presented to the pilot on his instrument panel in terms of speed or altitude. The pitot tube has been in use in aviation right from the early days, but today, with modern high-speed and high-altitude aeroplanes, other duties have been added to its role. For example, to

126

give a pilot a warning of an impending stall or loss of lift on the wings, the tube is associated with a stall detector and, as a result, a warning light can appear before the pilot, and additionally on some aeroplanes, movable nose flaps on the leading edges of the wings will lower to modify the wing characteristics to combat the onset of the stall. Our machine had such a detector and nose flaps. The pilot of the modern aeroplane, flying at the higher speeds and heights, refers to a machmeter rather than to his airspeed indicator. In these newer flight regimes, aeroplanes tend to nose up or down with changes of speed and to relieve the pilot of continually retrimming the aeroplane; it is done for him automatically with a device called an auto-mach trimmer, which receives its information from the pitot tube. The trimmer causes the elevator to make small corrections independent of the pilot's main demand. Again our bomber was fitted with such a device.

The auto-mach trimmer and the nose flaps had a common link – the pitot tube, and it was, in fact, the starboard tube that supplied the intelligence to these items. Another tube on the port wing tip fed information to the left-hand pilot's instruments and played no part in these other activities. The starboard tube supplied the right-hand pilot (co-pilot) with information, through his instruments.

Interestingly, all those items in the short list of features emerging from the wreckage examination had a common link. What would happen then if that common link failed, by becoming detached in flight? As far as the co-pilot's instruments were concerned, his machmeter and airspeed indicator would indicate virtually zero speeds. The altimeter would also register low. The auto-mach trimmer would have also received a spurious low-speed signal and promptly lowered the elevators to increase the speed of the aeroplane. At the same time, the stall detector would have told those nose flaps to extend.

Here was an aeroplane then flying at high speed, suddenly having low speed devices going into action, without the pilot's knowledge. Even if the pilots both looked at their instruments together, they would have seen disparity. But which is right? Are we flying fast or slow? It would take time to think, and that could be too late in modern flying machines. It seemed as though we did have the ingredients for disaster, but how could it have come about, and would the aeroplane have been in circumstances to exploit the situation?

The general flight plan for the aeroplane, as we understood, was to include manoeuvres that would induce shuddering and buffeting, severely shaking the pilots but not otherwise harmful to the aeroplane. It was also known that, at the altitude and speed for the manoeuvre, that aeroplane would have had marginal control. It must be remembered that this was a test flight, and normal day-to-day flying of these bombers never approaches anything like these conditions. In addition to the pilots receiving a shaking, so too would that starboard pitot tube, way out on the wingtip. It seems that the tube would have objected to such treatment, for extensive tests after the accident showed that the tube could have vibrated loose to the point of detachment, given the right environment.

Following this violent manoeuvre and loss of pitot tube and all that was consequential upon that, it was thought that the aeroplane would have nosed over and, being a very clean and aerodynamically efficient shape, would have built up speed to a stage where the pilots would have had extreme difficulty in effecting recovery of control. The result would have been inevitable: half a million pieces of wreckage to be sorted out at Farnborough.

During the investigation the fishermen were flown to Farnborough to see what was happening to all that debris they were pulling from the sea. What they saw convinced them that their efforts had been worthwhile. In turn, knowing what conditions they had been working under had spurred us on, not to waste the material and to glean all we could from it.

In the end, once again wreckage analysis had triumphed. No amount of armchair detective work could have dreamed up such a story. Truth is often stranger than fiction.

# CHAPTER 12

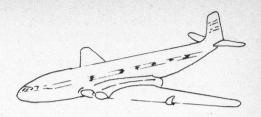

# Such a Little Screw

British European Airways Comet 4b G-ARJM flew out from London Airport on 21 December 1961, calling at Rome, Athens and Istanbul. At Istanbul Captain Ruddleston and his crew, who had spent the previous night at Istanbul, took over the aeroplane to fly to Ankara, in Turkey, Nicosia and Tel Aviv.

The take-off from Istanbul, the flight to Ankara and the landing at Ankara were all normal. The turn-round time at Esenboga Airport, Ankara, was forty-six minutes – that is, between landing and the start-up of engines for the next flight. During the whole period on the ground snow was falling, and the flight engineer supervised the sweeping of the top surfaces of the wings and elevators.

The recording of R/T conversation between the control tower at Ankara and G-ARJM showed that all was well and that the aeroplane taxied out by the short taxiway and backtracked along the runway to its take-off position on Runway 21, at the intersection with the longest taxiway. The runway length available was reduced at the time by some 3,300 feet, leaving 9,029 feet. The surface wind was calm; it was still snowing, and visibility was about $1\frac{1}{4}$ miles.

The take-off, at 9.30 p.m. GMT, was witnessed by the No. 1 Controller in the Tower, and by a BEA Senior Traffic Officer, who made a habit of watching all Corporation aeroplanes take off. Both men agreed that the take-off run and the lift-off from the runway were normal and that the aeroplane then rapidly assumed an abnormally steep climb. They were slightly at variance in their recollections of the stage at which a wing dropped or dipped, and of the sound of the engines. They did agree on the steep descent in a very flat attitude, right down to ground impact. Both men were in ideal positions for observations – one in the darkened control tower, the other outside on the roof of the terminal building, at a distance of about 400 yards from the runway, at its closest point. They could see the aeroplane – not just its lights – and both described its descent and ground-impact attitude in a way which lined up with marks on the ground. These observers had just seen a huge four-engined airliner rear up into the night sky, stall and fall back to the ground. The aeroplane was destroyed and many people were killed, although there were some survivors.

One of the survivors, the Israeli Air Attaché at Ankara, said, the afternoon after the accident, that after unstick the aeroplane climbed away very steeply at, he thought, forty to forty-five degrees. At 300–400 feet the right wing dropped and levelled, and then the left wing dropped. There was a reduction in power, and the aeroplane began to drop. He said that during the descent power came on again but not very positively. Another survivor, an American who had boarded the aeroplane at Istanbul and had not flown in a Comet aeroplane before, described the take-off from Istanbul and the landing at Ankara as 'beautiful'. The take-off at Ankara, though, was very much steeper.

A runway inspection was made after the accident, but nothing abnormal was seen or found. At the time of the accident the captain of the aeroplane had flown Comets for 785 hours and his two co-pilots for 537 and 961 hours.

An official Inquiry Commission was set up by the Turkish Government, and the Accidents Investigation Branch sent out an accredited representative, with other members, to assist if required. BEA also had people with the Commission, but as observers. The technical aspects of the Inquiry, including the wreckage investigation, were, however, delegated by the Turkish authorities to the accredited representative from the UK. This work was then done by a team drawn from BEA, De

Havillands, the manufacturer and the Accidents Investigation Branch.

There were no immediate indications of any simple cause for the accident, but in view of the event having occurred virtually wholly in the pitching plane of the aeroplane, main attention was being centred on the elevator controls and associated systems. Initial investigations were made at Esenboga, and all parts were then to be returned to the UK. It was not anticipated that there would be any quick results.

On 10 January 1962 the AIB made a request for assistance from the aerodynamicists of the Royal Aircraft Establishment, with regard to the aerodynamic effects of snow covering the top surfaces of the tailplanes, and what effect, if any, this could have on the climbing angle of the aeroplane.

On Thursday 11 January 1962. I made an examination of the tiny filament lamps used as warning lights on the Comet, and determined that the stall and ice warnings had been illuminated when the aeroplane impacted the ground. I did some further studies of these lamps on the following Monday.

On Wednesday 17 January 1962, together with my Divisional Head, I visited the AIB headquarters in London and had a general discussion on the accident. We had previously had some talk by telephone, and in consequence some wreckage items were already arriving at Farnborough, whilst we talked in London.

The next day, Thursday, we went to Hatfield, where members of the AIB, BEA and De Havillands, who formed the investigation team, were holding a periodic meeting to discuss progress and consider evidence and information already in hand. We went along to further our background knowledge before I started my own work on the wreckage. It was planned that on the following Monday the first of the main sections of wreckage would be brought back to Farnborough. In fact, an airlift by our Beverley aeroplane was planned.

During the meeting at Hatfield many possibilities for causing the violent pitching up of the aeroplane were put forward and discussed at some length. I said that I might be able to glean some knowledge from the instruments and so, as a starter, it was agreed that I would give priority to cockpit instruments and inverters when I started my examinations. The

131

inverters featured because I would be able to show their operating state at impact, and if for any reason they were not functioning, the instruments in turn would have been deprived of electrical power; in some way, the meeting felt, the pilot might have been presented with an erroneous picture by the instrument panel before him. Remember, the take-off had been at night, so instruments would certainly have been vitally important to its safe execution.

On Friday 19 January I set to and examined the inverters. There were three, No. 1, No. 2 and the stand-by unit. I had previously devised a simple technique for studying these items, back on Britannia G-ANCA, that aeroplane with the auto-pilot story in 1957. Simple cooling fans were keyed onto the shafts of the inverters, and the fan blades could be readily trapped during impact, and the shaft keys broken and sheared. It was an infallible check – study the shaft keys. I was able to show that both No. 1 and No. 2 inverters were OK with broken keys, and that the stand-by could not have been in use. In other words, power supplies to the instruments had been available to the bitter end.

Next day, Saturday, Jimmy Lett of the AIB who had done the 'fieldwork' at Ankara, came down to Farnborough, to debrief and identify the material already there. The task was completed by lunchtime and he departed for the weekend. Since I had food with me, I decided to seize the opportunity of a few quiet hours, Saturday afternoon, to look over instruments etc., thereby easing my workload when the masses of wreckage started to arrive next Monday.

I ate my lunch, then decided to look at the cockpit instruments. I had before me instruments for the pilot and the co-pilot, including altimeters, beam compasses and director horizons. (These latter instruments showed the aeroplane's attitude with respect to the horizon.) I started at random, picking up each instrument in turn, giving it a general examination, noting serial numbers, state of damage and so on. As the instruments were moved in my hands, the pointers moved freely within the instruments – this was all quite normal. The instruments had not been damaged to a degree to prevent this happening.

About the third instrument I picked up was the captain's

director horizon, as I turned it over, I noted that the pitch pointer was moving freely up and down over the face of the instrument, but whilst it went to the bottom of its travel quite normally, it only went a little way above centre in the upwards travel – and then stopped each time, as though something was obstructing its path. The outside of the case was quite undamaged, and there were no dents or distortions such that part of the case would have moved inward into the path of the pointer. I immediately referred to the co-pilot's director horizon, and the pitch pointer moved normally all the way up and down. I looked at other instruments that would have been next, or near to, the director horizons in the cockpit; all were in order. What was wrong with the captain's instrument?

At this moment I was simply curious about the pointer – anything to do with the accident itself was not in my mind. I was just being inquisitive – situation normal.

Having made notes about the instrument's condition, I carefully broke the manufacturer's seal, undid some screws and withdrew the instrument from its case. This simply meant removing a metal box covering most of the instrument but left that part of the face and pointers still enclosed behind the bezel glass. Thus the instrument's interior had now been revealed to me but the vital mechanisms were still protected from direct handling or touching.

Here I must explain that the pitch pointer was a thin bar across the face of the instrument. It turned at right angles to go along the side of the instrument to the rear, where it was pivoted in a moving coil mechanism. The long arm which moved up and down the side of the instrument was called the spider.

Within seconds of my removing the instrument casing, my curiosity was satisfied. A small screw, used to hold part of the instrument face in place, had loosened several turns, and its head had thus moved outwards into the path of the spider of the pitch pointer, such that the pointer could move downwards to its limit of travel, but as it moved upward, above the central position, the spider came into contact with the screw head.

My brain went into automatic overdrive immediately. The instrument had not been damaged in the crash, and no crash forces could conceivably have jarred or loosened that screw enough turns to place it as I had found it. So it had to be like that before the crash – and to have loosened progressively.

I had been noting down a sort of blow-by-blow account of my actions and findings – a bit like a bomb-disposal man relating his actions in case something goes wrong. I did it because I had to record every action. Once disturbed, my subject could not necessarily be replaced exactly as before. I realized afterwards that my note-taking, which was quite instinctive, had down these words: 'Christ – this must be it.'

During take-off the pilot eases the control column back, and the nose of the aeroplane rises. As it does so, the pitch pointer moves slowly up across the instrument face, faithfully telling the pilot the angle his aeroplane is at. The pilot knows that when he has achieved a certain angle the aeroplane is climbing out at a required speed and he then holds the controls steady until his next set of actions is required. With that screw in the way, the pointer had to stop. And then what happened? The pilot would have eased back the control column and the pointer moved appropriately upwards and then stopped. The pilot would have had no indications that this was an instrument fault and would have continued the back pressure on the control column. Of course the inevitable would have happened – just as the witnesses say – a steep climb, a stall and a crash.

On just one or two occasions in my career I have felt an inner sense of elation – a sort of tingling and excitement – this was one of those occasions.

I immediately stopped and locked the instrument away. I rang my Divisional Head and one of my staff to join me on Sunday morning, to examine the instrument independently. (I purposely avoided mention of the screw), and then we would discuss the matter when they too had seen it.

I went home that evening with my thoughts going round and round the problem. Surely it could not be this easy?

On Sunday morning my two colleagues made their separate and independent examination. They too spotted the screw, and they too could see the implications. We promptly phoned the accredited representative at AIB and, by joint effort and a few hectic hours, managed to contact all interested parties for a session in my office on Monday afternoon.

Present were representatives from De Havillands, the Air Registration Board, Smith's Instruments (manufacturer of the instrument), BEA, BOAC (also Comet operators) and the AIB. I went through my actions of Saturday, revealed the screw for all to see and then, together with the two Smith's representatives, whom I knew very well, we carefully dismantled further, just sufficiently to answer further questions that had arisen. By the end of the afternoon all were accepting the evidence and departed, except the men from Smith's, who stayed on for an evening session with me, probing deeper. They had to accept the evidence of a loosened screw, but now they wanted to see if they had to put their house in order, as regards production methods. I had been the first person to take the case off the instrument since it had been assembled during manufacture, and the method of locking the screw was prominent in our lone discussions. Finally they too departed, accepting, but not happily, the evidence unveiled during the afternoon.

Tuesday was spent studying the instrument under the microscope, tying up loose ends and preparing my case for presentation formally to the wreckage investigation team on Wednesday. At that meeting, in addition to the AIB, BEA and De Havilland representatives, were Accident Investigators from BOAC and two senior technical men from the Air Registration Board. The meeting started at 10.30 a.m., and my item was the only one on the Agenda. Within an hour all had accepted not only my evidence but also that this was the most probable cause of the accident. Arrangements for sending the wreckage to me were cancelled, and I was asked to prepare a formal report for transmission to the Turkish authorities.

The screw that had caused all the trouble was so tiny – 0.2365 inches long and 0.065 inches in diameter, but sufficient, nevertheless, to lose an airliner and many lives.

# CHAPTER 13

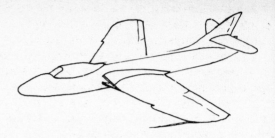

# *The Trees Give the Clue*

On 24 June 1964, on my arrival at my office, I was confronted with an atmosphere of urgency – would I go to the Empire Test Pilots' School immediately? In my absence yesterday, one of the school aeroplanes had crashed near Haslemere and the pupil had been killed. The Board of Inquiry wished to link up with me over the investigation. I received a general briefing on the accident, together with a specific request that I carry out a technical investigation for the Board. The pupil had departed on a perfectly normal sortie and in fact was returning to Farnborough when the accident happened.

The then Chief Flying Instructor for the School was Wing Commander Stan Hubbard, a sturdy, tough-looking little man, full of enthusiasm for his work. He arranged to fly down to Haslemere immediately, and I joined him in a Scout helicopter. We flew down to the scene of the accident and spent some time surveying it all from above. The countryside was very hilly and thickly covered with conifers. The aeroplane had crossed a ridge of high ground, striking through the rooftops of outhouses at a poultry farm, and then cut a swath down the slope through the conifers. There were no large pieces of aeroplane to be seen from above. It had broken up and was partly hidden in the undergrowth. We landed in a clearing and made a short walk among the trail of debris. As it took only about a quarter of an hour to fly to Farnborough, we returned home for lunch and then afterwards made another brief aerial survey before devoting the whole afternoon to climbing and scrambling up the hillside, through trees and bracken.

Whilst I was walking around the site at Haslemere, memories came to me of other sites I had visited. Those memories were stirred by the atmosphere and scent of broken and smashed tree trunks – the pungent smell of freshly hewn pine. Nowhere else can one experience it.

I made general notes of the site and discussed the manner of salvage with the Royal Air Force crew despatched to remove the wreckage to Farnborough. They had their special problems, unique to each site, and always tried to lift the wreckage in a way that would minimize further damage and possible loss of evidence to me. I always found it worthwhile spending some time describing what I was going to do with the pieces and then, when they came to Farnborough, would show them our reconstructions. Their change in attitude was always interesting to see. No longer did they feel they were dealing with rubbish; they realized they were playing a vital role in accident investigation – which of course, they were.

Back at Farnborough, I laid plans for the reception and reconstruction of the wreckage in the laboratory. Next morning I met the Board of Inquiry again and reported on my visits to the accident site. I then requested the procedure that I had found worked very well with Boards of Inquiry – that they should proceed with their own inquiries, leaving me to make my own analysis. I adjudged, on this occasion, that it might take two to three weeks for the wreckage to be delivered, assembled and examined enough, at least for first thoughts, so how about the Board visiting me on Monday 13 July 1964? Of course, if a story broke before that, they would know immediately.

In the event, the Board did come to see me on the 13th. I well remember the attitude of the President. They had more or less wrapped it all up, and this visit was really a formality to see me and the wreckage. They would then await my formal statement and close the Inquiry. This puzzled me, because I knew a lot of things that the Board could not possibly know, so I prepared for a bit of theatricals now in the presentation of my evidence, which, although not yet complete, could not be altered but only enhanced by anything else to come. I could not resist, however, asking what their answer was, and the President, somewhat condescendingly, said, 'Oh a misread altimeter, and then the aeroplane hit the ground during its descent.' I smiled, because the Board were now in for a shock – indeed, a little later in the morning they adjourned till the next Monday and went away 'very disturbed'; in fact, they had to tear up their already written findings.

If only they had waited, because the technical evidence that I produced made an intriguing and interesting story. Of course, most Boards would have waited to hear the technical evidence before proceeding so far.

What had I found that so bothered those three gentlemen of the Board?

First I must give a brief description of the fuel system on the aeroplane. Four tanks were housed on each side of the aeroplane centre-line; two in the fuselage, one in each wing and one under each wing, as a drop tank. All were connected with suitable valves to ensure a smooth and adequate flow of fuel. Fuel was drawn from the front fuselage tanks, simultaneously, and replaced from the other tanks, in the stream, as it were, so that the drop tanks would empty first. Low-pressure booster pumps, in the front tanks, transferred the fuel to the engine system where an engine-driven pump pushed the fuel at high pressure to the burners. If for some reason the booster pumps were switched off, or failed, only fuel in the appropriate front tank would be available to the engine – in fact, 575 pounds of fuel. Recovery of the booster pump would immediately restore full supplies. If one of the valves between tanks had a transitory failure, such as icing up, a fuel-transfer warning light would advise the pilot, and he simply switched off the appropriate booster pump; later normal balanced supplies would be restored by recovery of the pump. At the same time, although it did not constitute an emergency, the pilots of the EPTS were required to advise the control tower that they had been given a fuel-transfer warning. During the accident flight of our aeroplane that had crashed at Haslemere, the pilot had reported a fuel-transfer warning, and it was presumed that he had taken the appropriate action.

In the days following the accident, I reconstructed the aeroplane on my laboratory floor. I also examined and personally dismantled the engine. Whenever physically possible, I always liked to do this so that I could see evidence at first hand. I dismantled this particular engine after all the staff had departed one evening.

When I left later, I knew that when the aeroplane had crashed its engine was flamed out – that is, rotating but not developing any power. I had determined this from the state of the rotating parts, and everywhere was literally stuffed with wood dust (sawdust). As the aeroplane had flown through the trees, the engine had ingested bits and pieces of timber and had minced it into dust. Most important, there could not have been any flames, because nowhere did I see charred or burnt dust. Around the turbine shroud ring unburnt dust was stuck to the ring. This particular piece of evidence was to prove vital – that is, the evidence of dust stuck to shroud ring.

I contacted a friend at Rolls Royce at East Kilbride in Scotland. He dealt with problems on their products, and when I explained that I had a low-revving engine at impact, and one with wood dust, particularly adhering to the turbine shroud ring, he checked at his end and came back with remarkable news. To extract the resin from the wood, to form an adhesive around the shroud ring, the ring had to be at a certain temperature. That temperature would be attained only thirty seconds after the engine had flamed out – they had done tests at Rolls, on an engine, measuring temperatures. We could rely on the information. He arranged to come down to see the engine for himself, but I could accept what I had seen and been told.

Other items of evidence now began to pop into place. The port booster pump was not rotating when damaged in the crash, but the starboard pump was operating normally. Once I had assembled the pieces of wreckage together on my floor, a startling fact emerged. The aeroplane was severely burned all down its port side and wing. The starboard side was unmarked by fire. I was now dealing with a case of fuel asymmetry. The fire had not occurred in the air but as a result of the crash.

The story emerged: there had been a fuel-transfer failure warning; the pilot had switched off the port booster pump; he had continued his sortie and then made a descent towards Farnborough. Unfortunately he had not restarted the port booster pump, and about half a minute before the ridge of high ground at Haslemere, which was 780 feet high, his engine flamed out. He had plenty of fuel in the port tanks but had emptied the starboard side of the aeroplane, plus the port front tank. I learned from Air Traffic Control that the pilot, during his descent, had let down to 2,000 feet and then levelled, and it was about this time that he would have run out of available fuel. If he had tried a relight, which was unsuccessful, there was no time for a second attempt before the crash – but of course he had not restarted that port pump, so could not relight the engine.

The Rolls-Royce representative came down the same day that the Board of Inquiry returned. Not only did they have my story again but confirmation from the best source possible. Five weeks had elapsed since the accident, and all was done. An oversight by the pupil and an unfortunate coincidence of high terrain on his approach back to Farnborough. For me, how fortunate that the aeroplane chose to smash into those conifers: that wood dust gave the vital clue. A somewhat chastened but nevertheless grateful Board departed. Once more wreckage analysis had won the day.

# CHAPTER 14

# Accidents and Incidents

Accident investigation is not always a long-drawn-out process; occasionally a new par is set for the course.

A Royal Air Force air sea rescue helicopter, flying low along the east coast, near Yarmouth, suddenly plunged into the sea. A main rotor blade was seen flying away shorewards. Arrangements were made for the wreckage to come to Farnborough, for here was a structural failure – this must really be probed.

A few days later the first lorry arrived, and there, resting on the other wreckage, was the blade that had become detached in the air. I walked round the lorry just once, saw the blade, looked at its fractured end and said, 'That's it. We do not need the rest.'

An excellent example of fatigue confronted me. In simple terms, the blade had been weakened by cracking progressively until it could no longer hold itself together as a helicopter blade; the fracture surfaces told it all.

We guessed that the time on the job had been ten seconds. Of course, it really took longer to resolve the whys and wherefores of the fatigue by examination under the microscope and discussion with the metallurgists, but the initial identification for the reason for the blade detachment had taken only those few seconds.

For once, even that follow-up aspect was short-lived. I took the broken end of the blade over to my good friend Peter

Forsyth, the metallurgist – in fact, probably the world's leading expert on fatigue of metal. I have spent many hours in discussion with him over the years, and our two Sections at Farnborough were always 'open house' to each other and to our staff. Many is the time when one of us has had a visitor, often from the Accidents Investigation Branch, and has then phoned the other and passed the problem over, because it would be apparent that the problem was not as thought by the Accidents man.

Our joint studies on the helicopter blade quickly showed that the fatigue had been initiated by a man scoring the blade surface with a metal tool, when a plastic one should have been used to clean a joint in the blade skinning.

One day I was summoned to the Island of Sylt, off the north German coast. There was an aerodrome on the island, and RAF squadrons, based in Germany, would fly up for 'fortnight's summer camp' and practise live firing at targets towed out over the sea – a wonderful camp, if the weather was right, because Sylt was renowned for its nude bathing beaches.

During July 1955 two RAF jet fighters flew out over the sea early one morning for an exercise. On their return, one of the pilots reported a 'fire warning light' in operation. His colleague flew over and checked his aeroplane from the rear. Sure enough, he could see smoke coming from the engine region. There was only one action for the pilot to take – abandon, but he could not swim, so he stayed with his aeroplane until, near to Sylt, he could be sure of a dry landing from a parachute descent. The beaches were not yet occupied for the day – fortunately, because the aeroplane struck and broke up into many pieces. Retrieval later, by ground crews from the aerodrome, became a popular chore, searching for pieces of real and sometime imaginary wreckage among the nude bathers – if the bather was of the right shape and sex.

A couple of hours later, two more pilots flew off to perform their exercise. Suddenly one of the pilots shouted 'Fire!' and abandoned his aeroplane immediately. So now there were two crashed machines, one in the sea.

An AIB investigator went to Sylt to assist the Board of Inquiry. In turn he sent a signal to Farnborough for assistance

– there were technical features on which he needed advice. The RAF positioned a Special Communications Squadron Anson aeroplane from Hendon to Blackbushe, for me, and on Monday I was driven up from Farnborough to join the pilot. Blackbushe had Customs facilities, and I was off to Sylt. We flew first to Valkenburg in Holland where the aeroplane was refuelled. We then flew on to Sylt.

I linked up with the AIB man at lunchtime and we then went along to the hangar housing the planes' wreckage: the first badly burned and broken, the second just broken – not a blister. There had been no fire on that aeroplane. Our only conclusions were that the pilot, in a high state of alertness, after a colleague's accident two hours before, had seen what he thought was a 'fire warning light' and was probably the red glass of the indicator glinting in the sun. Other pilots had experienced such phenomena, but, of course, not at a time when tension could be high and fire warnings the reason.

We settled down to study the wreckage of the first aeroplane, and the story emerged fairly quickly. When fire attacks an aeroplane, the flow of air produces flame or smoke patterns on the aeroplane structure and, interestingly, the path and direction of progression of the fire damage are clearly shown. This is what we found on our aeroplane.

First, a brief description of the relevant parts of the aeroplane. It was a single-engined jet plane, with a short fuselage set at the centre of the wings. Two booms extended back from the wings to support a tailplane at the rear, high and behind the fuselage. The pilot sat in that part of the fuselage forward of the wings. Behind him were housed the engine, fuel tanks and an armament bay. The main fuel tank was housed above the armament bay, and the fuselage below the bay constituted a large door hinged along one side, which, when lowered, gave access to the bay and the tank area. A vent pipe was fitted to the fuel tank top and was bent over to continue right down through the fuselage to the inner surface of the bay door, when in the closed position. To ensure that any vented fuel escaped directly overboard from the aeroplane, a rubber grommet was fitted into a hole in the bay door, below the line of the vent pipe. A small length of rubber piping extended upwards from the grommet, sufficiently to allow the vent pipe end to enter and be below the top of the rubber piping.

What we discovered from our wreckage and from other aeroplanes of the same type makes interesting reading. The fire pattern itself emanated from an ignition source aft of the forward end of the armament bay, but we were able to trace fuel-flow patterns from the vent pipe position on the bay door. Due to a mixture of air-flow directions inside and outside the fuselage, we learned that the fuel could travel *forward* through the bay. But why had the fuel been there at all?

When the aeroplane was rolled in flight, a not uncommon event for a fighter aeroplane, fuel would enter the top of the vent pipe, which, with the aeroplane inverted, formed a 'U', and it stopped there. When the roll was continued, the fuel flowed down the vent pipe and away through the grommet. We now found some interesting evidence, suitably 'frozen' for us by the fire. Our vent pipe had not been sitting inside the rubber piping on the grommet. Instead, it had squashed the rubber piping, due to misalignment, and fuel had spilled over the inside of the bay door. We examined other aeroplanes and found other cases of misaligned vent pipes. The pipes and bay door grommet were all arranged to mate together, blindly, when the door was closed.

We had solved the mystery, but obviously another accident could well occur, if nothing was done to remedy the situation. Such design changes as flared or funnel tops to that rubber piping on the grommet, or other ingenious ideas, would take time. We, or rather the fighter pilots, needed an immediate cure. I suddenly registered, at this moment, that it was still my first day at Sylt. What a lot had happened in a short time!

I then realized that there was a check that could be made easily and quickly, and it would cost nothing. It would be made by the pilot before he climbed into his aeroplane. If the check was positive, he had no fears of fire; if negative, a spot of maintenance would put things right again fairly quickly. The central hole through the grommet, into which the vent pipe should enter from the top, was of course readily accessible from below, and with the vent pipe correctly in position, its squared-off end could be reached and felt by the insertion of an extended index finger. No ledge around the inside of the rubber piping meant, simply, no vent piping in position.

And so my recommendation to the pilots of the two squadrons at Sylt that evening was met with ribald remarks in the bar, but of course also with gratitude. I simply said, 'Insert your index finger up the orifice of your aeroplane before flight.' A spate of cartoons promptly went up on the mess walls – of a nature not suited to general publication, but nevertheless the point was getting across.

That Monday, when I had arrived at Sylt, morale among the pilots was noticeably low, understandably so – they were flying an aeroplane with an unknown fire hazard. That evening, everyone was smiling again, and the pilots of the squadron that had lost the aeroplanes insisted on entertaining us in the bar after dinner – and dinner was quite early. On Tuesday evening the other squadron on camp, not to be outdone, gave us another memorable evening. We had completed an investigation successfully and quickly. Obviously, longer-term solutions by local redesign would be made, and to that end the manufacturer had sent out a representative to discuss the matter with me before I left Germany. I arrived back at Farnborough less than sixty hours after departing Blackbushe, and in the meantime had removed a source of the aviator's dreaded fear – fire.

A jet fighter at West Raynham in Norfolk had been on a live-firing sortie at nearby ranges when suddenly there was a bang and a crack, and the cockpit canopy shattered. At Farnborough, one of the Sections of the Structures Department dealt with strengths and problems of canopies and transparencies, and I received a call from its Head to fly next day with him to West Raynham to see the broken canopy.

The next morning, 11 June 1948, I cycled over to Flight Dispersal, where I met up with our pilot, Flight Lieutenant Clive Brook, an RAE pilot. He had been the victim of the piece, some three years earlier. The Mosquito was an excellent aeroplane, and pilots loved it, but it had one unfortunate trait: if the pilot pulled back the control column, the nose came up and the 'g' forces came on – quite naturally – but with the increase in 'g', the force to further ease the control column back was reduced. There was a grave danger of over-controlling and breaking up the aeroplane.

Aero Department came up with a possible solution. Fix a simple tab to the trailing edge of the elevator and attach a bob weight to it on an arm, such that the effects of 'g' would force the weight down, and in turn this would put the tab at an 'up' angle. Such tab movement would induce 'down' elevator movement, and this would reduce the 'g' on the whole aeroplane. How would the tab remain still when the aeroplane was not under 'g' loading? A simple spring was fitted to return the tab to neutral behind the elevator. All good stuff, except that springs are not good damping devices on their own. It did not really need a scientist to show what would happen, but experts of Structures Department at Farnborough had given their blessing to the scheme.

Clive Brook was the pilot for the trials in the air, and a young aerodynamicist was responsible for the flight observations. They made one run from north to south across the Hog's Back, midway between Farnham and Guildford in Surrey. Some rather strange shaking was experienced after Clive had initiated mild 'g' loading. They agreed to repeat the run in the opposite direction. The strange shaking was experienced again, but this time increased in magnitude, and soon the whole aeroplane was involved. Suddenly there was a bang and the aeroplane disintegrated. During all of this, Clive had indicated to the observer to abandon the Mosquito. The last he saw of the young aerodynamicist was when he leaned forward to clip on a chest-type parachute – observers used these so that their movements were not restricted in normal flight conditions. Then Clive found himself out of the aeroplane and hanging below an open parachute. He must have reacted instinctively, because he had no recollections of going through any emergency or parachute drill. The aerodynamicist unfortunately fell to the ground with an unopened parachute.

Of course what had happened was quite predictable. The tab had been able to flutter, being restrained only by a spring. Its movements had been transmitted to the elevator, which in turn had vibrated and shaken the tailplane, and so on. The aeroplane was simply shaken to pieces. When the pieces were reconstructed in our laboratory, the whole sequence became self-evident.

Clive was still at the RAE, and that day was to take us to Norfolk in Anson TX210.

The incident aeroplane was in one of the flight hangars, surrounded by trestles and platforms to aid in the examination. We heard the pilot's story at first hand and then moved over to the aeroplane. Two features were readily apparent. The cockpit canopy had been shattered, and remnants of the perspex transparency material were still in place around the frame. A door in the fuselage, just below and towards the rear of the cockpit on the starboard side, was open and badly distorted. The door opened upwards about hinges along its upper edge. We were told that the door, giving access to some hydraulic equipment, had been struck by the canopy pieces. I was not happy about this. I had great difficulty in visualizing pieces of perspex travelling out and down to strike the door.

As we looked and talked, we were in the company of Air Commodore David Atcherley, the Commandant, and a retinue of officers ranging down to a lowly Flying Officer. They were from various sections at West Raynham and from Atcherley's own staff. All were interested to hear what the people from Farnborough had to say.

I suddenly had a thought – that door was still all wrong. I asked for the door of another aeroplane to be opened and, under the pretext of simply moving it open and shut, was actually noting the location of the forward lower corner, which of course would be forward and upper and nearest to the canopy when the door was open. I said nothing but returned to the incident aeroplane.

I told my Farnborough colleague that I had an idea, put my finger to my lips and, watched by the group of officers, mounted the platform to reach down into the cockpit. I then asked if anyone had searched inside the cockpit for pieces of perspex – no, surely it would all have been blown outwards and away? I excused myself and climbed up so that I hung head downwards inside the cockpit, and then I looked and felt around. I found what I had expected and made a quick glance at the pieces before surfacing from the cockpit. They bore the evidence I suspected might be there.

I then held the pieces up to the remnants on the canopy

147

framework and found that I could match the edges together. A pattern of silver paint appeared across the pieces. I pointed this out to David Atcherley, announcing that in my opinion the canopy had not shattered as a primary feature but had been struck by the door that had suddenly swung open. I was able to point out the shape of the forward lower corner of the door, in the paint smears. I then moved over again to the other aeroplane. I marked on the canopy the location of the paint smears and then, before the assembled officers, opened the unlocked door, and bingo – the corner matched the smears' position.

One comment was made by Atcherley which sufficed and in effect said, 'I agree and thank you' – it was 'Bloody Sherlock Holmes', and we all withdrew to the Officers' Mess for lunch.

To complete the story, examination of the door showed that it must have been incorrectly locked before the incident flight, and vibrations during the gun-firing exercise had caused it to open, rotate upwards and strike the canopy.

# CHAPTER 15

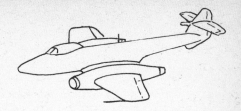

# Smaller than a Grain of Salt

At five minutes past two on Tuesday 11 March 1965, I glanced towards the north out of my office window in time to see a distant aeroplane plunge towards the ground. Minutes later I learned that an Empire Test Pilots' School Meteor twin-engined jet aeroplane had crashed at Cove and its pilot had been killed.

I promptly phoned the School Commandant and offered my assistance. A Board of Inquiry had yet to be convened, but I had acted on several occasions as 'guardian of the investigation' to protect a Board's interests, especially at the accident site, until formalities caught up and somebody was officially appointed. I did so on this occasion with the Commandant's blessing.

I straightaway made my way out to the accident site. I drove through Cove, and located the site. It was about on the downwind leg for the 07 runway at Farnborough. I had to park my car along a rough farm track and make my way on foot across a couple of fields to a point where a stream wound its way through the countryside. There was no trail of wreckage, and all of the aeroplane was found at the scene. It had fallen into the stream in a corner of a field used for grazing. The stream was narrow, with steep banks, and there was a sharp bend at the point where the aeroplane had struck. The land was boggy and the aeroplane had fallen on the inner bank of the stream. I found it partly submerged and partly on

149

the bank. Much of the wreckage was discovered only after probing and dredging in the stream.

Rescue and fire services were much in evidence when I arrived, still searching for, and finally recovering, the body of the pilot. Aware that the wreckage would be studied later, in great detail, the rescue personnel from Farnborough minimized disturbance of it in their efforts, although they had a difficult and thankless task in any event.

Next morning, at the site, I met a member of the Accidents Investigation Branch, who had been notified by routine Accident Signal of the event. I also met the members of the newly appointed Board of Inquiry. The AIB Inspector, after a look round the site and the half-submerged wreckage, said, 'It is all yours' and withdrew from the scene. I then discussed matters with the Board members and agreed to carry out the wreckage analysis for them, and any other associated technical matters. I estimated that it would take about three weeks to have the wreckage recovered, sent to my laboratory, reconstructed and examined, before I might have some answers for the Board. We looked around the site and then the rain came. Suddenly I had lost the Board.

Saturday afternoon found me in Wellingtons, squelching and splashing around that now very boggy mess. I was busy looking and making notes of all that I could see. That wreckage, inevitably, must be disturbed during salvage, because of its situation in the stream, so I had to have a clear picture before the move, to compare with what I would see afterwards.

Sunday was not a rest day – I visited my laboratory and chalked out a full-scale plan of the aeroplane on the floor.

Monday morning found me back at the stream. The distribution and condition of the wreckage showed that the aeroplane had descended nearly vertically, but slipping and skidding to starboard and with the nose well down. There was no indication that the aeroplane had been spinning or turning sharply at the moment of impact. It had broken up completely, but I considered that the forward speed had been low and that the aeroplane had not dived into the stream but had skidded and slipped sideways and downwards. The starboard wing and the forward fuselage, containing the cockpits (this was a

150

two-seat tandem aeroplane), had borne the main brunt of the initial impact. All of the extremities of the aeroplane could be identified at the scene, as were all the control surfaces and their tabs. The cockpit hood was present, and the pilot had been cut out of his harness by the rescue team.

Limited fire had occurred and was confined largely to the wing/fuselage centre section area which had landed on the inner bank of the stream. Fuel had flowed down onto the rear cockpit from the fuselage tank in the centre section, and much of its contents had been burned. A minor explosion had occurred after the aeroplane had crashed, and fuel had scattered downstream to burn local areas of the starboard wing and engine, which had become detached. The trees and banking were also damaged by fire for about thirty yards.

I could see that the fire had been caused at the time of the crash and the pattern could only be associated with the wreckage, as distributed locally at the scene. I took compass bearings on various local features and was able to determine that the aeroplane had struck the ground whilst slipping to starboard along a track 280 degrees magnetic, and with the nose pointing towards 195 degrees magnetic – that is, the aeroplane was facing across the general downwind trace of the 07 runway at Farnborough and had slipped sideways along the track.

I spent Monday, Tuesday and part of Wednesday out at Cove. Latterly I was working with the salvage crew, advising how they could best serve my interests. The first load of wreckage arrived in my laboratory during Wednesday morning, and to minimize further handling, the wreckage was then unloaded direct into its appropriate position on my chalked layout on the floor. By the end of Wednesday all was at Farnborough.

I started work on the wreckage in earnest on Thursday 18 March, just one week after the accident. The Meteor had been a sleek, silver, twin-engined jet plane, with a high tailplane perched near the top of the vertical fin. It was now several tons of muddy, twisted and broken metal. I completed the rebuilding and began systematically to study and analyse the evidence. By Wednesday 31 March I was in a position to say that at the instant of impact the aeroplane had been structurally intact and complete; its undercarriage had been down, and its flaps were up; the airbrakes had been in, elevator trim slightly nose up and rudder trim neutral. I could further show that electrical power had been available and so had hydraulic power. As far as the engine fuel system was concerned, the port and starboard low-pressure fuel cocks were open, as were the high-pressure cocks. The main fuel tank balance cock was shut. I knew that the port and starboard throttle levers had been at full throttle selection, and the throttle mechanisms at the engines and the throttle valves all confirmed this power demand.

I had found no pre-crash defect or failure in the flying control circuits or tabgear. The fuel contents gauges had been registering the fact that there had been adequate fuel available. My examination had shown that the port engine had been developing high power, with revolutions at the 10,000 r.p.m. or more. A very different story had been seen on the starboard engine. It had *not* been developing high power; in fact, its general condition had suggested a relatively low rotational state, with possible flame-out. It had certainly not been producing any effective power when the aeroplane had struck the ground.

I considered it very significant that, despite the symmetric high-power selections made in the cockpit, only the port engine was in the appropriate condition when the aeroplane crashed. I therefore decided to subject the starboard engine to further detailed studies. I found that it had been in good mechanical condition and that all drives to accessories and to the high-pressure fuel pump had been intact. I examined the pump itself and could see that it was still in good order and contained some fuel.

Thus far I had an aeroplane with nothing wrong, except that the starboard engine was not in accord with the pilot's demand from the cockpit, though the engine itself had been in good condition up until the crash. I now turned my attention to the fuel system, which in the Meteor with the Derwent engine was quite straightforward, comprising fuel tanks and low-pressure pumps; low-pressure cocks, high-pressure cocks and pumps; throttles and pressurizing valves and barometric pressure control units.

152

I could not determine the quantity state in the rear main fuel tank (the starboard engine supply), but I knew there must have been some fuel present from my examinations at the crash site. Unfortunately the low-pressure pump located in the base of the tank had been too badly damaged by fire for me to assess whether it had been rotating at the time of the crash. All the cocks for the system to the starboard engine had been at open and the throttle set full open; thus, if the fuel in the tank had been adequate, it would seem that the engine had been receiving fuel. If there was a problem in that fuel system, I was narrowing down the suspects.

I now turned to an item called the barometric pressure control unit. This is a unit which controls the fuel-pump delivery pressure. It is fairly simple, consisting of a capsule chamber and valve housing, separated by a rubber-bonded flexible metal diaphragm which acts as a pivot for a rocker arm. Fuel-pump servo pressure acts on the end of the rocker arm through a half-ball valve, which is spring loaded in the closed position; fuel-pump delivery pressure opposes the valve spring loading through a small diaphragm, piston and operating needle. This mechanism provides the control for maintaining fuel-pump delivery pressure at the fuel flow demanded by the pilot's throttle lever. Thus the high-pressure fuel pump is controlled by the pilot's throttle and the barometric pressure control unit half-ball valve, and the ball valve is critical in maintaining pump output appropriate to engine requirements. Any enforced opening of the valve – that is, lifting of the half ball from its seat – results in reducing the delivery pressure from the pump and consequently affects the engine performance.

I knew of at least one case where a foreign object two thousandths of an inch high had propped open the half-ball valve in a BPC unit and had caused the engine to flame out. I realized too that, depending on the size and time spent under the half ball by an object, the engine could have a momentary r.p.m. drop or, as already mentioned, a complete flame-out.

I now made my examination in detail of the starboard barometric pressure control unit. I could see that it was in sound mechanical condition up to the moment of the crash. The capsule was still airtight. I carefully stripped the valve chamber, putting it under the microscope with every move I made. There, between the half ball and the valve orifice surface, I found some foreign matter. I was able to study it under the microscope and measure it as well. It appeared to be a conglomeration of small particles, not just a single piece. It measured two thousandths of an inch by $3\frac{1}{2}$ thousandths of an inch by two thousandths of an inch thick, the latter being the amount by which the ball was being held off its seat.

I immediately contacted the engine manufacturer and gave a detailed account of all that I had done, and seen, and we came to a very firm conclusion that not only would the foreign matter have been sufficient to have caused the half ball to be held off and provide enforced bleeding of servo pressure but that the presence of the obstruction, and the state of the starboard engine, could not be accepted as only a coincidence. Together with the engine manufacturer, I had to conclude that the ineffectiveness of the starboard engine could be attributed directly to the presence of the obstruction.

I now had to fit my findings from the wreckage into the broader picture of the accident flight itself. I had learned from the Board of Inquiry that the pilot was carrying out a series of circuits and landings on the Meteor which included asymmetric landings and overshoots. The aeroplane had crashed on the downwind leg of the landing circuit. I had found an asymmetric power situation, but not one of the pilot's making. He had demanded full power from both engines.

That obstruction could only be introduced when the half ball was lifted on demand, and that demand would arise from pilot's selection of throttle towards closure or from capsule changes. The latter would not have been applicable and relevant to the flight conditions just before the accident. Thus the opportunity for the introduction of the foreign matter would have been limited to the last occasion when the pilot demanded throttle closure. The asymmetric manoeuvres could have been practised with either engine reduced in power but it was usual to use the port side as the 'failed' engine, because the starboard engine acts as a Master and supplies some of the services on the aeroplane. Thus the circuits at Farnborough on 11 March would have been made with the port engine being shut down at the appropriate times.

The general practice for a circuit was for the pilot to reduce power on the port engine on the downwind leg – that is, as he was proceeding parallel to the runway but in the opposite direction for the landing. He would then turn back towards the runway for the single-engined landing, or overshoot. As the aeroplane climbed away, in the case of the overshoot, the pilot would bring back power to the 'dead' engine, to fly up and away round the circuit at full power. Once on the downwind leg again, power would be reduced on both engines to allow level flight at circuit speed, until the point was reached to reduce power again on the port engine for the next landing.

In discussion with the Board of Inquiry we were able to marry its information and my evidence into a probable sequence of events. From a previous circuit, and now on the downwind leg again, the pilot reduces power for the level flight stage. *This would be the opportunity for the half ball to become obstructed.* The pilot is now proceeding downwind, preparatory to reducing power on the port engine but, unknown to him, the starboard engine is 'running down'. He now makes his power reduction on the port side and is confronted with a completely new situation. He is losing airspeed and height and is at a critical height for such a condition. The pilot's immediate action is to open up both engines to full power with his throttle levers, but, of course, only the port side could respond. Thus the aeroplane is subjected to increasing asymmetry and yaws and rolls to starboard and quickly moves into an uncontrollable condition, descending to the ground. These were essentially the conclusions reached by the Board of Inquiry, with which I fully concurred.

It remained for me to submit formally my detailed statement of findings to the Board. The problem of resolving the whys and wherefores of such an obstruction building up in the fuel system became a subject for further investigation between the manufacturer and myself. Ironically, the system was fitted with very fine filters, but this obstruction was built up of particles which conglomerated after passing through the filter.

My final action in this accident case was to appear before the local coroner's court at the inquest into the death of the pilot. I had one small problem: the coroner and his jury would be laymen, as far as aeroplane fuel systems were concerned. How could I put over the minuteness of the size of the obstruction? I took a pinch of salt to my laboratory and selected a grain and put it beside the obstruction under the microscope. *The obstruction was smaller than a grain of salt.* That would surely capture the imagination of the coroner and jury. It did. Two days later the local paper carried the story of the inquest under the banner headlines 'GRAIN OF SALT CRASHES JET PLANE'! But seriously, the efforts in the boggy field and laboratory had once again proved the value of wreckage analysis.

# CHAPTER 16

# *Fuji Strikes*

During the troubles in Malaya in 1965, an RAF helicopter crashed in Borneo. Parts of the machine were specially flown to me for examination for the Board of Inquiry. In consequence, in March 1966 I found myself on an aeroplane bound for Singapore to appear as an Expert Witness at an RAF court martial. The accident had been attributed to negligence on the part of a ground engineer and he was facing charges of negligence.

I returned to Britain at midnight on Friday 25 March 1966, just ten days after departing for the Far East. I had flown fifty hours in Britannia airliners, been through the mysteries of a Service court martial, experienced equatorial conditions in contrast to the snows I had left behind in Britain, and all because of my examination of a few small parts of the crashed helicopter.

On Tuesday 29 March 1966, my second day back, Group Captain John Veal, then Chief Inspector of Accidents, rang up to talk about the British Overseas Airways Corporation airliner, Boeing 707, G-APFE, which had crashed near Mount Fuji in Japan.

When an aeroplane like this crashes, the investigation rests with the country in which the crash occurs, so the Japanese carried out the investigation, with representatives of the aeroplane and engine manufacturers and of the country in which the aeroplane was registered and operated, all invited as

157

observers. The Japanese could call on these representatives for help, guidance and advice if necessary or desired. The Fuji accident was being investigated by a formal Japanese Commission of Inquiry, and the representatives including the FAA, NTSB, Boeing from America, and the Accidents Investigation Branch, BOAC and Rolls Royce from Britain.

John Veal's call was simple and straightforward. There was a request for me to make an independent examination of the wreckage in Japan. We came to an excellent arrangement. One of his investigators, Jimmy Lett, would go ahead of me to check that the wreckage had been collected and laid out in a suitable manner. He would also make his own studies. He would in effect filter the masses of evidence and information. I would then be able to make a specific attack on particular areas so highlighted. The wreckage was already being collected, and delivery to the reconstruction building would begin on Monday 4 April 1966.

In the event, Friday 15 April saw me on my way to London Airport once again, just three weeks after returning from Singapore. Some had observed before my departure that I was setting out on something like a twenty-four hour flight on the very type of aeroplane that I was to examine in Japan, which had broken up in the air. To be quite honest, the thought had not arisen in my mind at all.

When we landed at Tokyo, I met up with Jimmy and Gilbert Jamieson, also of the AIB, Ben Folliard, Chief Inspector of Accidents BOAC and Ted Church, an Inspector from the Accidents Investigation Branch. We were to be joined a few days later by Don Hellyer, one of Ben's inspectors.

On the Monday morning, after a day's rest to catch up on lost sleep, we travelled out to Chofu aerodrome. The wreckage of the 707 had been collected in a large wooden hangar, used during World War II to house Japanese fighters for the defence of Tokyo. The hangar was filled with debris, but I could recognize the general layout of wings, fuselage and tail unit. An interesting feature that I had not seen before in wreckage now confronted me: small bunches of flowers among the broken and twisted remains of the 707. I learned from Jimmy that these had all been placed on the wreckage, along with joss sticks, out at the accident site at Gotemba, by the local

158

population. Flowers had been placed wherever a person had been found dead, and 124 persons had died. A sobering thought. Ten years later I was to see a somewhat similar feature in Argentina when I visited the scene of the accident to an HS748 airliner.

Before continuing with the story of Chofu, I must give a brief account of the accident.

BOAC 707, G-APFE, which was operating Flight BA911 from Tokyo to Hong Kong, crashed on the south-east side of Mount Fuji, near Tarabo, fifty-five miles west of Tokyo, at 14.15 hours, Tokyo time, on 5 March 1966. All on board had been killed.

The aeroplane had taken off from Tokyo at 13.58 hours and was cleared for a VMC climb, via Fuji, to join the airway at Rebel. The weather was fine between Tokyo and Fuji with no cloud and good visibility. After routine communications associated with the departure were completed, when the aeroplane was climbing through 2,000 feet, nothing further was heard from it. At about 14.15 hours a large number of witnesses saw the aeroplane falling from a considerable height, trailing white vapour. A number of witnesses claimed to have seen the aeroplane breaking up in the air, at a height which was later calculated from eyewitnesses and from the photographic evidence as about 13,800 feet, near to Mount Fuji. Fuji is 12,400 feet high. The wreckage was strewn over a distance of 10 miles.

Investigators from the AIB and BOAC arrived on the night of 6 March and left for the scene early on 7 March to make a field examination. The Japanese Ministry of Transport had appointed a Commission to investigate the accident. The Commission included Professor Marizos from Tokyo University as the President, and other professors, Japanese airline officials and members of the Japanese Civil Aviation Bureau.

Arrangements were made for the wreckage to be transferred to the American air base at Chofu, near Tokyo. The arrangements also included, at this stage, the assistance of two AIB Inspectors (Jimmy and Gilbert), myself from Farnborough and Boeing accident investigators. Thus, when we entered the hangar on Monday morning, 18 April 1966, the wreckage had been assembled and examined by Boeing and AIB investigators and now awaited my efforts.

I was shown the salient features already established and then left to make my own examinations. In parallel with examining the wreckage, I also undertook another task – the calculation of trajectories for the separate pieces of wreckage as they fell to the ground. I did this back at the hotel at Tachikawa. Thus we had a routine worked out for the next few days: Chofu during the day and Tachikawa in the evening – wreckage then trajectory calculations.

On 27 April, nine days after I had commenced my examinations of the wreck of the 707, I presented my findings to the Japanese Commission. Both British and American parties were present in a Conference Room in Chofu. The Commission members were ranged along one side of a long, wide table, and we all sat opposite. I was centrally placed opposite to the President, Tomigino Marizos.

I had done my homework carefully for this occasion. It was to me, personally, a very important day. I had been invited to fly across the world and make an independent analysis of aeroplane wreckage for the Japanese Commission, and now was the moment of reckoning. I had made examinations and sequences of failures and presented them to Boards of Inquiries many times in the past, but the Japanese Commission was an unknown quantity. I wrote out an eight-page script in large writing, spelling out in the simplest terms what I had done, how I had done it, why and what I had achieved.

At the meeting with the Commission, a Japanese interpreter stood close by my left shoulder, and I held my script so that he could easily read it as well. I had not had time to prepare a copy for him. As I went through my statement, my words were simultaneously translated. I was certain that most of the Commission understood me perfectly, but all their questions or replies came in Japanese.

After making appropriate greetings to the assembly, I described my efforts. The aeroplane had broken up extensively in the air, but my interests had centred upon the detachment of the right wing; detachment of all engines and their nacelles; detachment of the forward section of the fuselage; detachment of the vertical fin, and detachment of the horizontal stabilizers – the tail surfaces.

Aerial view of accident site of delta research aeroplane at Blackbushe, Hampshire, 1949.

Aeroplane wreckage illuminated in trees only by its own entry path.

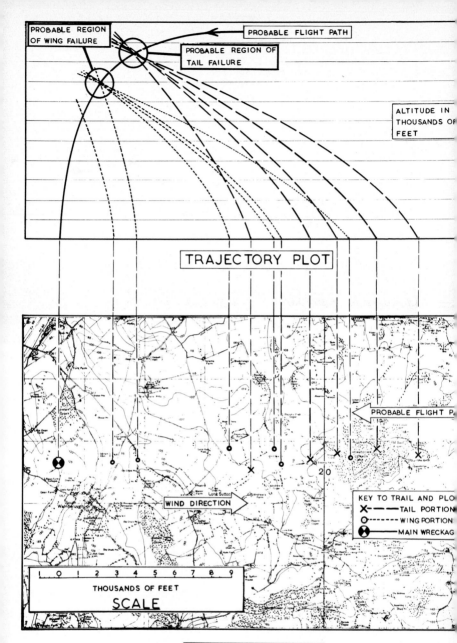

Example of a wreckage trail and trajectory plots to determine height of disintegration after an aeroplane has broken up in the air.

Not wrapped for delivery – collapsed wing over gondola at crash site.
'Bunty' inflatable machine.

Reinflated and tilted on wing tip for investigation purposes.

Accident to RCAF Otter VC3666, April 1956. Artist's impression of actual break-up in the air, based on the author's analysis of the wreckage and sequences derived therefrom. (Drawing by Bob Bradford).

*Opposite top*: ETPS Meteor crashes into a stream during a training flight from Farnborough, March 1965.

*Opposite below*: Author studies barometric pressure control unit from starboard engine of Meteor under microscope, and finds evidence of obstructed valve.

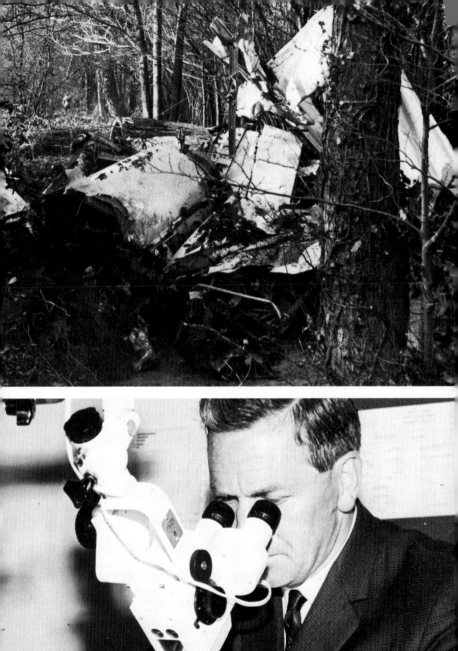
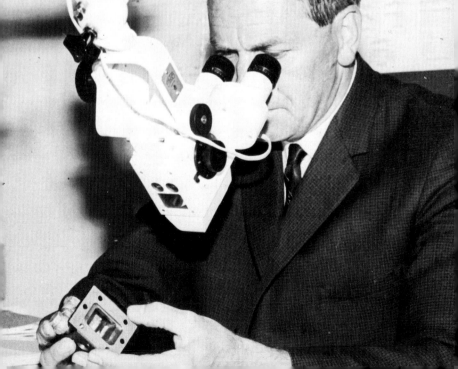

The broken BOAC 707 airliner G-APFE descends, March 1966. Not fire but fuel vapour.

Main wreckage of the 707 in Japan.

Wreckage of the 707 at Chofu, Japan, for examination.

Author examines tail unit of 707 at Chofu.

Accident site of the Queen's Flight Whirlwind helicopter, December 1967.
Note detached rotor blades at bottom left of picture with rest of the
helicopter burnt out at the top of the picture.

Fractured main
rotor shaft from
Queen's Flight
helicopter,
evidence of
fatigue clearly
visible around
left-hand
portion of
shaft.

I was able to tell the Commission that the right wing had failed by bending upwards, the failure area embracing the No. 4 (right outer) engine position – a significant feature. I had seen no evidence of any pre-crash weakness or failure that could have contributed to the wing failure. All of the engine nacelles had failed in the same manner, by moving nose to the left, and because of this common feature, and the fact that No. 4 engine was located in the area of the right-wing failure, I could only conclude that the engines broke away before the wing failure. My examination showed that the forward section of the fuselage – that section forward of the wings – had become detached by bending to the left – that is, in a similar manner to the detachment of the engines. The vertical fin became detached by moving top to left, to fold down and strike against the left-side horizontal stabilizer and cause its breakage. This breakage had led to the detachment of the right-side horizontal stabilizer – the whole horizontal stabilizer being initially a single slab of structure right across the rear end of the aeroplane. I could conclude then that all tail-unit damage and detachment had followed failure of the fin.

I was able to conclude that in all probability the fin failure and the tail unit damage had occurred before the forward section of the fuselage had failed. A careful study of the control circuits for the horizontal stabilizer assisted me in this matter.

I was unable to find any physical link of evidence between engines and fin, in sequence, but since all engines had failed in the same manner and the forward fuselage too, it was extremely likely that all had been subjected to a common circumstance which could have occurred if the fin became detached, and, of course, I could also show the local sequence between forward fuselage and tail unit, through the control circuit study.

My study of the wreckage of the Boeing 707 led me to conclude that the fin had been the first item to become detached in the air. One could picture an aeroplane skidding violently round with the fin off, suddenly presenting its right side to the airflow and the engines, fuselage and right wing all then receiving an abnormal loading to bring about their failures.

Following my detailing of my findings from the wreckage

examination, I moved on to the matter of trajectory calculations. I explained that the method I used in Britain took into account the wind drift characteristics of the pieces of wreckage. I also chose pieces with low terminal velocities which would lose all forward speed as they left the aeroplane. The pieces then fell under the effect of gravity and drifted in the wind. I assumed a drag coefficient of unity and from experience decided the manner in which each piece would fall, either flat or edge on, and so on.

I explained to the Commission that I had dealt with many structural breakages in the air and had found that I could usually ignore the local effects of turbulence and other weather phenomena.

I pointed out that the calculations I had made for the Boeing 707 accident near Mount Fuji were not intended to give exact sequences of breakage or of heights but that the plots of the trajectories, or paths down to the ground, did give a general indication that the aeroplane could have broken up first at a height between 14,000 and 18,000 feet and that the orders of separation were in agreement with my conclusions from the examination of the wreckage. The calculations showed that the wing was not the first thing to detach and that the forward fuselage detachment was late in the sequence of events.

I then told the Japanese that I had concluded that the fin was the first thing to fall. I could not formally, at this stage, offer any explanation as to why it had done so, since this would require consideration of loading data and information by people more qualified in that sort of work. However, my personal thoughts, not relayed formally to the Japanese at this stage, were that the aeroplane had encountered very turbulent conditions as it was flying downwind of Mount Fuji. We had discovered that the aeroplane was not flying the regular route out to Hong Kong but more likely had diverted to give the passengers, many of whom were Americans, a close-up of the famous mountain.

In conclusion I thanked the Commission for the opportunity to examine the wreckage and said that I would be interested to hear what its own investigation would conclude on this matter.

The meeting then continued with the members of the

Commission firing questions at me, through the interpreter, concerning the sequence of failure and the trajectory calculations. The meeting then closed and I could see that it had been very successful.

The next morning we resumed our visit to the wreckage hangar to tidy up outstanding matters, and suddenly, at about mid morning, I received a visit from three learned professors, who had been sent specially to learn all about trajectories from me. This was to me particularly gratifying, not only because here were three very clever gentlemen anxious to see me but also because the Japanese are not usually so openly demonstrative of a lack in their knowledge – because of the inevitable 'loss of face'.

The next few days were spent in collecting details and consolidating information, for preparation of my report back at Farnborough.

The twenty-one days I spent in Japan were hard work but tremendously interesting and exciting. We had visited what in some respects might be thought to be a somewhat alien country; we had completed our task successfully and we were leaving, having established excellent relationships with our hosts. A small part of our role when overseas is, I suppose, as a sort of ambassador, and I like to think that on this occasion we came away with the flag flying.

# CHAPTER 17

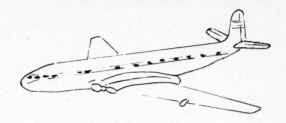

# *People Supply the Clues*

Comet G-ARCO left Heathrow Airport on 11 October 1967 as Flight BE284, bound for Athens in Greece. On board were thirty-eight passengers, three flight crew and cabin staff. The flight was uneventful, and the aeroplane arrived at Athens at 01.11 hours on 12 October. Six passengers for Athens disembarked and twenty-seven new passengers joined the flight, now relabelled Cyprus Airways CY284 and bound for Nicosia.

The aeroplane's transit at Athens was quite normal: it was refuelled, and baggage for the passengers joining the flight at Athens, together with newly boarded freight, was placed in holds No. 1 and 2. Baggage and freight from London for Nicosia remained in holds No. 4 and 5. The aeroplane became airborne at 02.31 hours and fifteen minutes later reported crossing R19B, an airway reporting position, at 29,000 feet. At 02.58 hours G-ARCO passed an outbound Comet of BEA which was flying at 28,000 feet. Each aeroplane saw the other, and the captain of the westbound Comet later indicated that flight conditions were clear and smooth.

R/T communications were being recorded at Nicosia, and it is known that G-ARCO called to establish contact at 03.18 hours, plus nine seconds. The message to Nicosia Control was simply: 'Bealine Golf Alpha Romeo Charlie Osc …'. The end of the message was clipped. It has been established that the Comet was then at position 35°17′ and 30°17′, about fifteen miles to the east of R19C. Efforts were made to contact G-ARCO, but without success, and overdue action was therefore taken. The aeroplane had been in the air forty-seven minutes since leaving Athens.

A search and rescue aeroplane of the Royal Air Force left Akrotiri, Cyprus, at 04.40 hours, and $1\frac{3}{4}$ hours later, whilst at 1,000 feet altitude, sighted the wreckage of G-ARCO, in the vicinity of R19C. The crew of the search aeroplane described the scene below generally as a kidney-shaped area of flotsam, bodies and oil slick. There appeared to be two areas, the smaller one to the north. The search was now made in runs at 50 to 100 feet above sea level.

Just over an hour later, another search aeroplane arrived over the scene, at an altitude of 2,000 feet, and its crew also saw a kidney-shaped fuel slick 500 by 200 yards, with flotsam in it. On investigation, at lower level, it became obvious to the crew that the flotsam was in two areas. A secondary area was found one to $1\frac{1}{2}$ miles to the south of the fuel slick, where items such as blankets, seat cushions and inflated life-jackets were seen.

Location and fixes for the scene of the tragedy, obtained from aeroplanes and surface craft engaged on the search and recovery, showed the area to be about twenty miles from the last radio message position and, in fact, back along the original track of G-ARCO. Several vessels were involved in rescue efforts, but ultimately all recovered bodies of victims were concentrated on two ships. Nineteen were taken in a Turkish ship to Turkey and later airlifted to Rhodes, thirty-one bodies were transferred to a Greek naval vessel and were taken direct to Rhodes. By and large, the victims had been in the water for nine to ten hours before recovery.

To personnel on the search at sea level, the situation became much more grim and poignant. The second coxswain of one of the surface rescue vessels was concerned with the recovery of six bodies. His detailed descriptions, together with those later from pathologists, were to play a critical part in the investigation soon to get underway into the sudden loss of the Comet aeroplane. The coxswain was able to provide sex, age, state and degree of clothing, jewellery etc. and condition of body with reference to external and obvious injuries. Of great significance was the fact that everything was soaked in

165

AVTUR-aviation-kerosene fuel and that there was no sign of burning on anything he saw or picked up. He also confirmed that one life-jacket was still in its container and others were inflated, with straps dangling but not cut.

The surface craft on the recovery action were deployed in the two areas, 'northern' and 'southern', and all bodies and flotsam were identified in the first instance by these areas. The coxswain had made his recoveries in the southern area, and the investigators were later to note that all fuel-contaminated bodies and flotsam were from this area, whilst clean and uncontaminated bodies and flotsam were found in the northern area.

The post-mortem examinations by the pathologists were made under very difficult conditions and under great pressure to complete their task quickly, due to lack of refrigeration facilities.

It must be evident to most people that aviation accidents inevitably have violence associated with them – aeroplanes blowing apart in the air or falling thousands of feet to smash into land or sea. In this present case it became apparent to the pathologists that they were dealing with two general categories of injury: extreme – for example, 'very badly injured, being split from base of spine to stomach through the crutch, right side of buttocks sliced through, smashed left arm with biceps missing, found completely naked, floating face downward' to others virtually free of superficial injury and fully clothed. With a few exceptions, the extremely injured category were also fuel-contaminated, whereas the lesser injured were not contaminated.

A further division of material evidence occurred in the recovery programme itself. The northern area flotsam included part of the forward toilet of the aeroplane, life-jackets, personal belongings and a graviner fire-extinguisher bottle, normally located in the starboard centre section of the wing. The southern area flotsam was mainly cabin furnishings, seat cushions, carpets and parts of the galley. It also included some personal belongings, handbags, life-jackets and three life-cots, of the types provided for small children. Some of the life-jackets were inflated, some were uncased but not inflated, and some had remained in their containers. All of this material

166

from the sea was destined to appear in my laboratory.

The whole area of surface debris and oil slick was slowly drifting, and after four hours it was close to the airways reporting position R19C. This was determined by the RAF search aeroplanes, from radio bearings from Rhodes and visual bearings on prominent features of the Turkish coast. Later in the day, at 11.30 hours, the German ship *Astrid* fixed the position of the two areas by radar bearings and distance from the Turkish coast. By this time the area had also increased to 3.75 nautical miles in length. Taking into account the surface drift and these observations, it was estimated that the approximate position at which the aeroplane fell into the sea was 35°55′N and 30°01′E.

Flotsam from Comet G-ARCO was delivered to Farnborough on 17 October 1967, just five days after the accident. It was held in my laboratory for examination initially by the AIB, although I was able to make some superficial 'interest' examinations myself, without disturbing any of the material. Very quickly, however, my interests became firmer, as I was tasked with making formal examinations for the investigations.

I was dealing with a problem in which there was no actual aeroplane structure or components available, only contents and furnishings. Fortunately these items could be identified as to location on the aeroplane, and I could already see that, armed with such information as passenger complement on the flight from London to Athens and from Athens to Nicosia – to determine seating arrangements – and a knowledge of the injuries, contamination etc. of the bodies, some patterns could perhaps be produced and a 'form of aeroplane' be laid out on my laboratory floor.

The seating of the passengers could be obtained from passengers and cabin crew who disembarked from the aeroplane at Athens and from airport staff at Athens responsible for the passenger boarding on the accident flight. I considered it unlikely that any gross re-arranging of passengers would have taken place after take-off. One particular group of passengers who boarded the aeroplane at Athens were 'block booked' at the rear of the tourist cabin.

As a separate issue I received ten wrist-watches which had been recovered from the bodies of passengers. The watches had been stopped because of damaged hands and dials, displaced pinions and/or ingress of water. In all cases I was able, by detailed examination, to determine the time indicated on each watch when it stopped. I was able to conclude that the time of impact of the wearers of the watches with the sea, and hence the time the aeroplane struck the sea was about 03.25 GMT. I had noticed that two of the watches were dry inside but that the other eight contained traces of kerosene. In the three cases where the watch and wearer could be positively identified, the watch was contaminated and the wearer had been found in the southern area.

I was also considering the possibility that the aeroplane had broken into at least two major pieces in the air and I needed the weather situation at the approximate time of the accident, principally the wind strengths and directions, for use in calculations of falling items.

The flotsam received at Farnborough was identified and positioned on a full-scale layout of the aeroplane on the laboratory floor. All the material belonged to the fuselage and, like the bodies, could be divided into that which was contaminated by kerosene and that which was 'clean'. Interestingly, all items known to have been located forward of the transverse datum of the aeroplane, roughly near where the wing-leading edge joined the fuselage, were free of kerosene. All items to the rear were contaminated.

A mass of clothing had also been received, some identified, some not, to particular passengers. On the classification due to fuel contamination, the division into front and rear of the aeroplane still held good, where we knew a particular passenger's seat. Gradually pictures began to form as the days went by, and my analysis and discussion went something like this.

The AIB investigations had established times and positions of the aeroplane along its flight track. It could be seen that the flotsam and bodies were found nearly twenty miles back along the track from the estimated position of the last message. The natural division of the flotsam, after identification and layout on the floor, by state, recovery area and location on the aeroplane, had produced two groups, strongly indicating that the forward fuselage had landed in the sea in the northern area, and the rear fuselage and associated wings, containing the fuel tanks, had fallen in the southern area. Some seat cushions were found to the south between one and $1\frac{1}{2}$ miles from the southern area. I considered then that separation of the fuselage into at least two major portions must have occurred before the aeroplane struck the sea, to account for the distribution and state of the flotsam and bodies.

My examination of the flotsam suggested that the division of the fuselage, in the fore and aft sense, could have occurred at about the transverse datum (centre section, front spar). The study of the passenger injury pattern supported this finding. In general, passengers forward had sustained relatively slight or intermediate injuries, whereas those to the rear had sustained intermediate or extreme injuries. It was apparent to me that such a division could only arise from the passengers experiencing very different circumstances, following, say, the case of breakage of a fuselage in the air. I therefore made some trajectory calculations on the premise that a fuselage separation occurred in the air at the front spar position. The aeroplane design firm indicated to me also that this would be a likely point of separation under an ultimate loading condition.

I first of all plotted out some trajectories from a specific point at 29,000 feet. The resulting scatter of flotsam and bodies at sea level was far too large to be reconciled with anything seen by the first search aeroplanes over the scene. I then plotted trajectories, from sea level upwards, from the general positions of items, as suggested by charts and maps from the search aeroplanes. A very close interception area of plots was found at about 15,000 feet, so I had to conclude that the aeroplane had not broken up at its cruise altitude but at the lower altitude to produce the general pattern in the sea of flotsam and bodies and the damage and injury pattern to bodies. This conclusion was supported by the pathologist's report that there was no evidence of explosive decompression in any of the bodies, although this would have been expected if major disruption of the aeroplane had occurred at the higher altitude.

I had examined that lone fire bottle found in the northern area and had concluded that it had been discharged by operation of the fire-extinguishing system crash inertia switch and that this occurred before the aeroplane struck the sea. The operation of the crash switch could have been by inertia loading, if the forward fuselage became detached from the aeroplane under high positive accelerations. The inertia crash switch was located in the nose wheel bay, in the nose of the fuselage.

If the aeroplane had broken up at about 15,000 feet, the heavier pieces of wreckage, containing passengers, or the falling bodies of the passengers, would have taken about two minutes to have fallen to the sea. With the time of final impact deduced from the watches as 03.25 GMT, break-up of the aeroplane would have occurred at about 03.23 GMT. This is five minutes after the last recorded message from G-ARCO and would have allowed ample time for the aeroplane to have passed its last estimated position on track at 29,000 feet and returned to the position of break-up at 15,000 feet. I found no evidence, however, to give any indication of the flight path of the aeroplane during the descent from 29,000 feet to 15,000 feet or to explain the final break-up.

Independently of myself, Eric Newton of the AIB had been working away at other aspects of the investigation. He had examined the seat cushions and found some very interesting evidence. As a result, he linked up with an explosives expert, some experiments followed, and I received the following information.

An explosion had occurred on the aeroplane before it struck the sea. It had taken place just above the floor of the aeroplane, under the rear of a tourist-type seat. The seat, occupied at the time of the explosion, was one on the extreme port side of the aeroplane. One body, marked No. 6 for the investigation purposes, had been injured in a manner to suggest that it had been in a seat one row to the rear, and one seat to the right of that under which the explosion occurred. No other body, recovered, contained evidence of a comparable nature, to suggest close proximity to the explosion.

At the specific request of the AIB, I was now asked to attempt to locate the likely site of the explosion.

To recapitulate, it appears that the explosion occurred under a tourist seat on the left-hand side of a row of three, that the seat was occupied but the occupant was not recovered. Only No. 6 had injuries of the type expected from an

170

explosion, and it had been the occupant of the middle seat in the row behind. Unfortunately, No. 6 was one of the group of block seating, so that his precise location in the cabin was not known, although it would have been in line B in one of the rows 3, 4, 5, 6 or 7. The line A, port side (or window) seats in this area would also have been occupied by members of the group, but none of them was injured in a manner to suggest that they had been in close proximity to an explosion. However, one of the group was not recovered.

My analysis and considerations now proceeded thus. The nature of the damage to the seat, over the explosion, and to body No. 6, in line B, suggests that the effect of the explosion was not unidirectional. It is likely, then, that a person seated immediately to the rear of the explosion would also be injured by it. Since no such person has been recovered, it is possible that this was the person in Seat 4A, against a port window (known but not recovered) or was the missing group member who may have been seated in either 3A or 7A – that is, behind either of two known but not recovered persons. It would seem then that the explosion could have occurred under seat 4A, 5A or 8A, and that passenger No. 6 was in seat 3B, 4B or 7B.

Seat 8A was located over the centre section containing fuel tanks. At the stage of the flight reached, when the explosion was thought to have occurred, the tanks would have been empty of fuel but containing fuel vapour. Discussions with the explosives expert resulted in the conclusion that, had the explosion taken place under seat 8A, it was very likely that a major disruption of the aeroplane centre section would have followed. In this event, I considered it highly improbable that any of the patterns of recovery from the sea, injuries to passengers etc. which so clearly indicated a major separation of the fuselage, near to the front spar, could have been produced. I could only conclude, then, that the explosion was most likely to have occurred under either seat 4A or 5A and that passenger No. 6 was in seat 3B or 4B.

My overall conclusions from the work that I had done were as follows: the aeroplane struck the sea nearly seven minutes after making a last radio call to Nicosia, and in that time had descended from 29,000 feet to approximately 15,000 feet and then broke up into at least two major portions. I also concluded that the explosion, known to have occurred before the aeroplane struck the sea, was probably sited under either seat 4A or 5A, in the rear tourist cabin on the aeroplane, and that the effect of this explosion was not to disrupt the aeroplane catastrophically. These conclusions were accepted and formed part of the official accident report released on the accident.

# CHAPTER 18

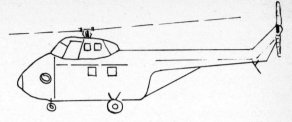

# *There Are No Exceptions*

A small number of military passenger-carrying aeroplanes and helicopters are held at RAF Benson, where they are serviced, maintained and kept in immaculate condition, inside and out. Where not painted, externally, the metal surfaces are usually highly burnished. The very high standard is achieved because these are aeroplanes and helicopters of the Queen's Flight.

Although specifically selected and kept for this purpose, the machines otherwise are not treated any differently from other Service aeroplanes and helicopters. Queen's Flight machines are instantly recognized by the immaculate scarlet finish and, except where they are actually carrying the Queen or members of the royal family, fly on routes as any other military machines. On the occasions of their carrying royalty, the route is usually designated a 'Purple Airway', and all other aeroplanes, military and civil, will be notified to keep the airspace clear.

Many flights are made by Queen's Flight machines without royal personages on board, when they are used for the Queen's business, as when, on Thursday 7 December 1967, one of the helicopters left RAF Benson to go to Yeovil. On board were four RAF officers, including Air Commodore John Hubert Lampier Blount, the Commodore of the Queen's Flight. The party was proceeding to Westlands Helicopters Limited at Yeovil, to discuss the purchase of helicopters for the Queen's Flight. The helicopter being used for the flight by the Air

Commodore was a Westland Whirlwind, a type well proven and much used by both the RAF and the Royal Navy.

At five minutes past nine on that morning, Bryan Peter Brown, a farmer at Leckhampstead, Berkshire, was standing beside his tractor, looking due north, when he noticed a helicopter flying from east to west, about $1\frac{1}{2}$ miles from him. Suddenly the complete set of rotor blades lifted off the helicopter and floated down like a huge sycamore seed. The machine itself plunged to the ground and burst into flames. Bryan hurried to the scene and saw that very little remained of the helicopter but that the rotor blade system, about twenty-five yards away, was virtually undamaged. There were no survivors.

Gilbert Jamieson of the Accidents Investigation Branch, who had been in Japan with me, went along to assist the Board of Inquiry, and during the day I had a telephone call from him, asking that I examine part of the helicopter urgently. An RAF Wessex flew into Farnborough next morning with the hub of the crashed helicopter. The rotor blades had been detached for convenience of transport, but apart from that the part was still as recovered from the crash site.

Before we touched the Whirlwind hub, it had to be sprayed. It had come direct from the crash site, as foot and mouth disease was prevalent in that area.

The hub is normally attached to the upper part of the rotor shaft, which extends upward from the main gearbox in the helicopter. In effect, the helicopter is suspended by the rotor blades on that rotor shaft. It was immediately apparent that the shaft had broken in flight, thereby releasing the rotor blades from the helicopter. Why had the shaft failed? That was why Gilbert had sent the part to me for closer examination. That closer examination told me one thing: the shaft had been grossly cracked prior to the separation which had caused the accident. The magic word 'fatigue' reared its head.

I contacted Peter Forsyth, the leading fatigue man at Farnborough, for his advice on the finer points of the failure, but it was also necessary that we see the rest of the shaft, which would contain the mating surfaces to those we should now see.

It is an interesting fact, that, when a failure occurs, the matching surfaces are not always identical as would be

174

expected – after all, they were one piece which simply separated; but during the act of separation, many local features can appear on only one piece which, if then taken out of context, can give a false picture. It is essential, then, to see both portions of a failure.

I contacted the Board of Inquiry and explained that we confirmed Gilbert Jamieson's suspicions but that it was essential for us to have the rest of the shaft. The wreckage of the crashed helicopter had been removed from the accident site and taken to RAF Odiham, a major RAF helicopter base in England. The shaft duly arrived from Odiham, having been dismantled from the gearbox. I took it immediately to the metallurgists.

On occasions like this, I practically lived in Peter's laboratory, watching and studying, discussing and arguing the pros and cons. We were always on the most excellent terms, and our respective laboratories and facilities were always available to each other whenever the need arose. Often my role here would be to interpret the metallurgists' findings for the Boards of Inquiries or investigators seeking advice from Farnborough. The reverse process often applied too: I would make examinations to determine whether there was a need to consult the metallurgists. Boards and investigators, not always conversant with the sort of features produced in air crashes, would innocently take any odd-looking feature to the metallurgist, and this usually out of context with the rest of the story. Peter and his staff could only pass judgement on what they were given. One of my roles was as a sort of technical interpreter on accident matters at Farnborough.

The story on the helicopter shaft emerged quite quickly. We worked throughout the weekend, because once fatigue had been discovered and confirmed, it was necessary to determine whether it was an isolated feature on the crashed helicopter or whether it could be present on any helicopter of that type.

That shaft, in the region where the failure had occurred, was about $5\frac{1}{2}$ inches in diameter, tubular, with a wall thickness of about half an inch. The metallurgists were able to tell me – and I could see for myself – that the origins, or starting points, were very similar at several points around the periphery of the shaft. The origins had been produced during manufacture of the shaft. To put a fine surface on the shaft, it was subjected to a grinding process, and it was now that local flaws were created, called 'grinding burns'. These flaws led to minute cracks. These cracks, of course, were not seen; nor was their existence even suspected at that stage of manufacture. Further manufacturing processes followed, the result of which, on the one hand unfortunately, assisted in the development of the cracks and on the other hand made the cracks even more difficult to detect, even if their existence had been known. As time went by, those minute cracks on that shaft had corrosion forming in them, and stress corrosion made them bigger. Gradually, as the helicopter was used, each origin developed into major cracks, progressing through the shaft material, and these in turn connected to each other. Finally, the stage was reached where insufficient whole material remained to support the helicopter and it failed completely.

Unfortunately all of this had taken place in an inaccessible part of the helicopter, and all normal inspections and servicings would have completely missed the cracking. No one associated with the helicopter, whether on the ground maintenance side or the flying side, would have had the slightest inkling that catastrophe was inevitably approaching. It so happened that disaster struck on a 'business' flight, sad though it was, and not on a royal flight, which it could well have done. The ramifications of that would certainly have been very far-reaching.

We knew that we had not discovered just an isolated type of failure. It could well be present on other helicopters. Immediately investigations were put to hand to determine such things as the batch of material from which the shaft was made; the machine on which it was ground, the operator – in fact everything that could help to determine if other cracked shafts could have existed. Its being in a normally inaccessible location on the helicopter meant that inspection of all other shafts called for major dismantling to enable the checks to be made. Hence the exercise first to try to isolate suspect shafts, to avoid unnecessary work. Whilst these actions were being taken, methods of detecting the cracks had to be devised.

This sort of thing always cropped up when we discovered a failure or feature much like that in our helicopter shaft.

Usually these matters now passed into the hands of other people and the manufacturer to resolve, to balance the technical problems involved with the operational requirements of the aeroplanes or helicopters to fulfil certain and sometimes vital duties. How long would it take to check out each aeroplane? Could a reasonable risk be taken to phase the inspection into the next convenient maintenance period? And so on. Of course in wartime the operational risks involved in flying against an enemy might call for the technical inquiry to take second place. Always the discussions and studies were long and searching but urgently pursued to ensure that all aeroplanes could be kept in the highest safety state at all times. It came to light in the studies of the helicopter shaft failure that the cracks in question could be detected only by a process known as nital etching, and it was only as a result of the accident that this process came into being as a crack detection method on these particular shafts. Prior to the accident, the manufacturer could not have found the minute cracks by normal detection methods.

On Tuesday, just five days after the accident, I was able to give a verbal statement on behalf of the Farnborough scientists' efforts to the Board of Inquiry and the RAF Flight Safety Directorate, so that the various consequential processes could be put in train. On Thursday I sent off my written statement. There remained only my attendance at the coroner's court later to complete my part in the story of the loss of a Queen's Flight helicopter.

# CHAPTER 19

# A Man's Worst Moment is Relived

The new helicopter was on a flying trial with a Royal Navy pilot and three Army personnel on board. It was flying low over the terrain when suddenly the pilot spotted a larger helicopter ahead. He made a banked turn to starboard, as evasive action, but shortly afterwards the new helicopter crashed and all on board, except the pilot, were killed. The machine was destroyed. A Board of Inquiry subsequently found that the helicopter had crashed out of control, following excessive control movement by the pilot. This had 'stalled' the hydraulic servo jacks, used to assist the pilot in altering the angles of the main rotor blades.

About six months later a similar helicopter, with two qualified instructors on board, was approaching its base to land when suddenly the control column started jerking to the left and forward and back again at a very regular frequency. It took both instructors' efforts on the control column to bring the helicopter under control and make an emergency landing. The instructors were unhurt, and the machine suffered only minor damage.

A Board of Inquiry was set up to investigate the incident, carried out extensive inquiries and tests and was able to show that the frequency of movement of the control column, as judged on test by the instructors from the incident, was the same as that of the occulting anti-collision lamp, mounted on the tail. The Board also determined that, in some way, the controls of the helicopter were being alternated between manual and hydraulic power operation.

To assist pilots of helicopters in their control of the rotor blade angles, or pitches, as they were called, hydraulic servo jacks are fitted into the control system, so that the pilot initiates a demand and the servo jack carries it out, having a much more positive effort capability. On the helicopter in question a small electrical solenoid valve could be moved into one of two positions at the touch of a switch, so that hydraulic power could be used; or the jacks could be locked and manual power could be applied by the pilot. For all normal flying, the controls were power operated but for emergency practice purposes, occasionally, the helicopter would be flown manually to accustom the pilot to the different feel or forces to be encountered on the control column if hydraulic power was ever lost in flight. This was always done with the helicopter flying steady and the pilot making the transition knowingly.

At this stage of the inquiries I was asked to accept the helicopter, to examine it to try to determine the nature of the fault or feature that had suddenly started that column oscillating. The helicopter was delivered by road, and before it was placed in my laboratory, the Board ran the engine to provide power to demonstrate their findings to me to date. Then the helicopter was defuelled, wheeled into the building and trestled into a flying attitude, so that I might test, strip and examine. The Engineering Member of the Board of Inquiry, Squadron Leader Trigg, joined me in my investigation.

First we acquired an independent power-supply source so that we could provide both electrical and hydraulic energy to the helicopter when required. We also incorporated an oscilloscope into the electrical circuit which supplied power to the solenoid valve. Thus we could see displayed the characteristics of the electrical energy actually reaching the solenoid. I had a closed-circuit television and camera with video recorder and set up the camera and microphone so that I could record the events in the helicopter cabin.

We also introduced a flasher unit, as used for the occulting anti-collision lamp (and very similar to that used for motor car direction indicators) into the solenoid circuit. The Board had already shown that this was the likely source to produce the appropriate frequencies experienced at the control column by the instructors.

One more refinement was added. We could not rotate the rotor head and blades to produce the air loads that would have to be reacted by the hydraulic servo jacks, but we could estimate the sort of loads to be expected, and simulate them in a simple manner. Trigg devised a lever system at the rotor head and, being of the appropriate bulk, sat on the lever at different arms' lengths from the blades to represent the air loading. A trial run, with all power sources activated, soon had him sitting at the right position.

I sat at the controls of the helicopter, sound and vision were switched on, one of my staff stood by the switches for power and flasher unit, and we were ready for our tests.

When the tests were in progress, I found myself fighting the controls to hold them central and stationary, just as the instructors had done. We went through a whole series of runs, and soon we felt we knew how the whole system worked and what effects could be created by altering various features. I played back the video recording and found we had an excellent display of the likely events in that helicopter when the instructors had experienced the control oscillations. In a way, we had been doing my 'getting to know the wreckage' phase of an investigation. We could now consider how the problem could have arisen, and make further tests and examinations to try to unearth the fault or feature responsible.

After one of our runs, the Squadron Leader and I, in almost the same breath, said, 'How about Rod's accident?' That was the fatal accident, some six months earlier. I sat in the helicopter again and strapped myself in, whilst he disappeared on top again, remarking that he had an idea. He then relayed a message down to me to apply certain control settings, as though I was actually reliving the moments before the accident.

The pilot has two main controls for flying a helicopter. By his left side is a lever which he can raise or lower, rather like a handbrake lever in a motor car. This lever, called the collective pitch control, does just that. It collectively, or simultaneously, alters the angles of all the main rotor blades, thereby increasing or decreasing the overall lift of the main rotor system. In front of him, and looking just like a control column in any aeroplane, is the cyclic pitch control. By moving this sideways or back and forth, the pilot can tilt the helicopter to the left or right, or move it forward or backward. I set the pitch lever in my left hand, at about half travel upwards, and held my hand steady. With my right hand I held the control column slightly back and to the right, as though I was asking the helicopter to tilt to the right and nose upward, just as the pilot would have done to avoid the larger helicopter ahead of him.

I sat there waiting. Suddenly I felt the control column try to move forward and over to the left, and at the same time that lever in my left hand started to lower itself. I held on tight, trying to return the controls to their original settings. I treated all of this very seriously, as though I was really flying that machine, and my life depended upon it. Soon I was straining with all my effort, and just succeeding. Then, as I felt that I would have to release the controls, Trigg called, 'Switch off.' The controls suddenly slackened, and I slumped, my face feeling hot with the exertions.

He climbed down from the rotor head, came round to the cabin and saw me for the first time. He had no idea what had happened, and laughed, at first, to see me somewhat distressed. When I explained that I had been fighting the controls to hold them steady, he became very excited, and so too did I when I learned what he had done. It was so simple. He had simply switched the controls from power to manual – that was all. After I had recovered from my efforts, we repeated the exercise, and again I had my fight, with the whole event being recorded. Could it be that, whereas we were looking for a means of 'oscillating' the solenoid valve, for the instructors' incident we also only needed a simple fault that would switch the solenoid valve to suddenly to steady manual, to explain the accident?

After some discussion we contacted one of the investigators of the accident, to invite him to Farnborough and also to bring with him, if possible, the surviving pilot. In the first instance, he came alone. He saw me fighting the controls and experienced it for himself, so he immediately arranged for the

181

pilot to come to Farnborough the next day. He did not, however, tell the pilot the real reason for the visit, simply that we wished to talk about his accident.

When they arrived, we talked over the pilot's experience together, and I explained that we had the incident machine here to investigate the strange control phenomena. Since we had introduced electrical and hydraulic power for our tests, and thereby had a 'working' helicopter, perhaps he would like to sit in it with me and talk his way through his accident flight. He strapped himself in and played with the controls for a spell, until he felt that we had got them set, as though he had been flying just before the accident, I then asked him to move the controls into the positions into which he had put them to avoid that other helicopter. He co-operated fully, and just as we reached the critical condition, I silently signalled out of my door and watched the pilot closely. I saw him tense up as he felt the controls move away from him; then he was fighting just as I had been. After about twenty seconds, I signalled 'stop'. All this time, the camera and microphone had been recording the pilot's efforts.

His first words after the test were, 'What was that? It was very like my nasty, apart from being slightly lower in the magnitude of the forces than those I had to apply.' We simply explained that we had only switched to manual. Would he mind if we did it again? This time, we introduced higher loads for the pilot to react. At the critical moment, I again called for the switch to manual, and again I saw a fight with the controls, but this time his efforts were far more intense. I let him continue for about twenty seconds again and then halted the test. Rather breathlessly, the pilot said that the test had been somewhat nearer the accident conditions, from his recollections. I noticed that he was rubbing his thumb towards its base, near to the index finger. I could see that he had a reddened area, where I too now had a bruising from my tests.

He said now that the only thing missing in the 'replay' was the actual position that the column had attained during his fight. We made the necessary adjustments, and I gave him time to relax. Then he was strapped in again – willingly – and we were off once more. Again I had him suddenly fighting the controls, but this time I did not stop the test. Eventually he just gave up, let go the controls and slumped. I waited for him to recover – but I noticed that he was gazing out through the roof of the helicopter. He slowly turned to me and, in a broken voice that was quiet and quavering, simply said, 'That was it – that is what happened to me.' It had been unnecessary really for him to have said that, because I could see with my own eyes that he had just relived a traumatic experience.

Remembering his thumb and the rubbing, I pointed out that I had similar bruising, but then he surprised us. There on his thumb was an old scar, now healing. He had really fought the helicopter on the accident flight. The bruising and marking had come from a control button at the top of the control column, and when under pressure, the hand gripped the column and the button was forced into the thumb.

The next morning I contacted one of the leading psychologists at the Institute of Aviation Medicine, and played through my tape of the pilot and his fight. The psychologist said immediately, 'You have done that pilot a great service. You have obviously made him relive completely his traumatic experience. You handled him perfectly.'

We had still to find out why the helicopters had been 'biting their pilots', so we settled down to the slow, laborious task of carefully dismantling the electrical wiring on our helicopter.

As with all modern flying machines, the electrical circuits can be multitudinous, and wires can be routed everywhere. For ease of handling, neatness and convenience of maintenance at later dates, the wires are collected together in bundles, called looms, and are strapped together with plastic straps. The loom or pre-formed harness is then easily installed, just like a large, single, fat wire, with various branches leading off to the appropriate components. The plastic straps are wrapped around the loom, and the free end passes through a slot at the opposite end of the strap. That free end is inserted into a sort of hand gun which, when fired, draws the free end further into the gun, tightening the strap around the loom. A desired tightness allows the loom to be firmly gripped, but not cut into by the strap, otherwise the wires around the outside of the loom could have their insulation cut and the conductor wire within exposed.

All wires on aeroplanes and helicopters are coded, so that a required circuit can be readily traced or extracted. I now started to examine the loom with a specific thought in mind. If the solenoid valve had to be actuated at the frequency appropriate to that of the occulting lamp, ought not the two circuits have been brought together somewhere, so that the signals to occult the lamp would operate the solenoid? Would the appropriate wires be adjacent in the looms and would there be the opportunity for current to be directly transferred or induced from one to the other? A study of the electrical wiring diagram showed which of the wires could be expected to produce the desired results if brought into contact, and I found that they were in close proximity in the loom.

Carefully the loom straps were eased away, and immediately, at one position, we found evidence of broken insulation. The breakage had been caused by the edge of the strap. Several other wires were also found damaged, including some relevant to our investigation. We could not complete the picture in every respect but had sufficient evidence now to show what must have happened. In the case of the instructors' incident, the lamp or flasher wires had been able to feed the signals to the solenoid. In the case of the accident, it was a matter of cancelling the demanded solenoid 'On' signal, but, of course, without the wiring for examination, we could go no further.

How could the straps have cut into the insulation of the wiring? The guns used to secure the straps could be pre-set for different tensions and loom diameters. We made some enquiries and learned that, when works inspectors checked the tautness of the finished looms, they found them slightly slack. This meant a re-work by operatives and, because this occurred more than once, the operatives simply reset the guns to a more than adequate tension to save themselves extra work. The result: the insulation suffered.

And the accident pilot? Was the slate wiped clean for him now that a more likely explanation had been found? This sort of matter was out of our jurisdiction, of course, but there was the satisfaction that we had shown him that he could not have been responsible. He could now live with a peace of mind.

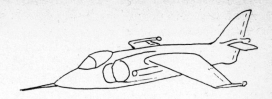

# CHAPTER 20

# *An Interest in VTOL Aeroplanes*

On 21 October 1960 an event took place at Dunsfold Aerodrome, Surrey, that would have far-reaching effects on future military fighting tactics. At that time, such a phrase as 'viffing'* would have been unknown – not even invented. That event was the first lifting off vertically of an aeroplane – not a test vehicle – under its own power. The aeroplane was the Hawker P1127, No. XP831, and that take-off had been restrained by tethering cables for safety.

About one month later, on 29 November, after being stripped of all extraneous equipment, XP831 made free hovering flight for the first time. 831 was later joined by XP836, and these two aeroplanes, during a series of trials, made successful transitions from vertical to horizontal flight.

On 14 December 1961 XP936 was being flown in the West Country by Chief Test Pilot Bill Bedford when suddenly one of the cold nozzles became detached. Bill tried to make a precautionary landing at RNAS Yeovilton, but the machine got out of control and he had to eject at about 200 feet. The aeroplane was lost, but Bill was saved.

I now became interested in VTOL aeroplanes, as they were then known, and made an examination and study of the failed nozzle. First, briefly, the workings of the aeroplane. A

---

* 'Viffing' – the sudden stopping of an aeroplane in forward flight by use of the rotating nozzles.

conventional-type jet engine had been heavily modified so that it no longer issued its gases directly astern through a tailpipe. That tailpipe had been replaced by a bifurcated arrangement so that the gases were directed out to each side of the engine. At the end of each pipe, and just outside the aeroplane fuselage, was located a rotatable endpiece – a nozzle containing a cascade of horizontally placed blades, angled downwards. The gases now emerged and were deflected downwards, and, given sufficient energy, the aeroplane could have been caused to lift up. By rotating the nozzle through towards the rear, the aeroplane could then be translated from vertical to horizontal flight. To augment this arrangement, a similar pair of nozzles protruded from the compressor area of the engine near the front, but these issued 'cold air'. The rear nozzles were made substantial and of steel; the forward nozzles, lighter, were made of laminated glass-fibre cloth. It was one of these fibre nozzles that had broken away from Bill Bedford's aeroplane.

The broken pieces were studied at the manufacturer's facilities at Kingston, and soon we were delving into the mysteries of the construction of such material. Simply, it consisted of layers of the cloth alternating in direction for strength purposes, much like wood grain in plywood. The whole assembly was soaked in a resin which, when hardened, provided a tough, firm product. We now discovered some of the problems of using such a material. Those nozzles were shaped with curves and bends and twists, and the laying-up of the cloth laminations was obviously very critical. It was in this area that the nozzle problem had been found. Of course it was resolved, and the world was to hear more of the developed versions of these early VTOL machines. I took every opportunity, whilst at Kingston, to learn all about these new flying machines.

Development proceeded, and the P1127 aeroplanes became star turns at displays. When XP831, that first historic aeroplane, flew at the Paris Air Show in 1963, as Bill Bedford made a hovering turn before the spectators, preparatory to flying forward, the aeroplane suddenly dipped and drove downward and forward into the ground. But it settled, and after a few minutes the cockpit canopy opened and Bill emerged, unhurt. A Board of Inquiry was set up, and the aeroplane was returned to Kingston. On 19 June I was asked to join the Board and make the necessary examinations and analyses of the aeroplane.

We were greatly helped by films taken of the event, and one in particular, taken from the right viewpoint, showed quite distinctly the nozzles moving from vertical to horizontal far too quickly to have been pilot-controlled. The studies homed in on the control mechanisms for the nozzles. All were very massive affairs, with chains, sprockets and shafts, because they were required to do heavy duty. The design of the nozzles and their cascades, and the swirling action of the gases emerging from the engine, all caused the nozzles to want to rotate aft, from the vertical. Now this inherent tendency was turned to account in the design of the control mechanism. It was so arranged that it reacted the rotational tendency of the nozzles, and by controlling the degree of reaction, the nozzles could be allowed to drive themselves aft. The mechanisms were then used to full effect when returning the nozzles to the vertical position. Quite ingenious and effective – but what had gone wrong? The controlling 'brain' was a finely machined valve rotating in a finely machined body. Small particles of dirt had entered leading to a malfunction. As a result the nozzles took charge and simply rotated themselves aft, there being no reaction from the mechanism.

As part of the investigation, the air motor was taken to the engine-test beds at Patchway, Bristol, and fitted to a similar engine in a test chamber. All was revealed as we had determined. Various filtering and modifications soon resolved that problem.

By the end of 1967, the Service version of a VTOL aeroplane had appeared – the Harrier. From time to time when an accident occurred, usually the inevitable teething trouble of a new aeroplane, I was able to assist, particularly on the instrument side. Special miniaturized instruments were now coming in, including a percentage RPM gauge. I devised a simple yet foolproof technique for getting these gauges to tell me their reading for the instant of ground impact of the aeroplane. On one occasion, a Harrier was undergoing special engine tests when it suddenly fell to the ground. In the time available, it was unlikely that the engine could have run down further than what it was when the descent started. Could I help? In the first instance I had the RPM gauge sent to me. I made my examination and determined that it was showing 87·5 per cent. I sent off this information to the Board and then received a very interesting reply. The engine firm had made some calculations and, taking everything into account that day, for the aeroplane suddenly to descend as it did mean that the engine revolutions had dropped to – yes – 87.5 per cent. Very gratifying.

On Tuesday 28 January 1969 a Harrier was lost at Dunsfold whilst being flown by an American pilot during a conversion course. The aeroplane had lifted off and, whilst making the transition into forward flight, had rolled and plunged to the ground. The pilot had ejected, but too late, and was killed. The President of the Board of Inquiry was Wing Commander Jack Henderson, our Commander Flying at Farnborough. He telephoned to ask if I would look after the technical side for him and handle the wreckage aspects.

I well remember that morning when I arrived at Dunsfold. No time was wasted: 'Look at this' – I saw a colour film of the accident. I saw the aeroplane lift off, move towards the camera, then roll and turn away. Out shot an ejector seat, and the Harrier plunged into the ground. Fifteen seconds from beginning to end, but when that aeroplane struck the ground, the immense fireball that developed made me, spontaneously, shout out, 'Christ, am I supposed to do something with that lot?' It all looked so terribly impossible. Of course we rarely saw these scenes, but they were common where aeroplanes were burnt up.

After a day of discussions and a visit to the accident site out on the airfield, I departed to prepare for a long day out in the field. That evening, when I went to bed, the weather was fine and dry. The next morning it was snowing, and it was lying thick and fast. That day out on the open airfield in the driving wind was unforgettable.

I spent the next week at Dunsfold and then the wreckage was to be delivered to my laboratory for closer scrutiny. Meanwhile the Board had been doing its homework, and they had been able to discover that the aeroplane had been 'lost' during the transition from hover to forward flight at a time when control by the pilot would have been very marginal, particularly if the pilot was distracted or preoccupied suddenly. So far I had not seen anything untoward in the wreckage, but of course I had not yet started 'digging' inside the aeroplane. That would come at Farnborough.

On Monday morning, 3 February, I called into my office *en route* to Dunsfold, in time to receive a call from Dunsfold: 'Come quickly' – what I was told had me motoring as fast as possible. I went straight to the wreckage. It had been left, because nobody would interfere, whilst I was still making my studies. I opened up a small hatchway and peered inside with a torch and mirror.

What had happened was this: whilst I was out in that snow on Thursday, another Harrier, at another airfield, had been flying and its pilot had felt a 'twitch' on the aileron control. After landing, the technical staff had discovered that a large jubilee clip, used for joining the air ducting behind the pilot, had become detached from that ducting and had jammed against the aileron control cable nearby. I found exactly the same thing when I had looked into our wreckage. However, I warned the Board that I had to determine whether our case had been caused when the aeroplane had struck the ground. I spent the next hour going over and over the evidence before me. Finally I had to accept that our clip had been against the cable before the aeroplane crashed. It was physically impossible for it to have got into position at the attitude the aeroplane had struck the ground. I still had to determine how and why the clip had become detached, but we had now found a distraction for the pilot at a critical stage in his take-off. I now left matters with the Board, as I had already cleared the aeroplane in general terms. I doubted if my work back at Farnborough would do more than support what I had found.

Our efforts, with others, did quickly resolve the problem of that clip, once and for all, and the Harrier went on to greater things. In fact, the Harrier has demonstrated in no uncertain terms how it has developed into an excellent fighting machine by its actions in the South Atlantic against the Argentinians in 1982.

The Harriers went into service, and such accidents as did arise were dealt with by Service Boards of Inquiries assisted where necessary by the AIB and I was usually called upon them only to analyse some specific features arising. As in all other aspects of aviation, wreckage analysis had helped in the development of VTOL aeroplanes.

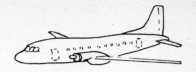

# CHAPTER 21

# Down Argentine Way

On 14 April 1976 the sound of aeroplane engines could be heard by Señor Humberto Yapura and Señor Livio Navarete, as they were taking their meal at the Yacimientos Petroliferous Fiscales drilling site at Meseta Buena Esperanza in the western Argentinian desert area. They looked up and debated the identity of the aeroplane. Was it an Otter or an Avro? It was a twin-engined airliner flying towards the south at an estimated height of 4,000 feet above ground level. They had just agreed that it was an Avro when suddenly, without any warning, the men saw pieces, including the starboard wing, become detached. The pieces fell to the ground only a few hundred yards away, and the aeroplane itself struck the ground about 6,000 feet from the oil drilling platform.

The aeroplane, LV-HHB, was of British design and manufacture, registered in Argentina and operated by YPF, the National Oil Company. It was making a staff transfer flight and had flown from Cutral Co in Neuquen Province, leaving at 14.00 hours and landing on a YPF strip at Rincón de los Sauces some forty minutes later. It was during the return flight to Cutral Co, with a crew of three and thirty-one passengers, that the disaster occurred. The employees at the drilling rig rushed to the scene in all vehicles available – motor trucks, shooting brakes, scrapers – for they knew that the aeroplane had been carrying friends, colleagues and relatives. Their efforts were in vain, for all the occupants were killed instantly. A tragic loss

191

for all present and for the Company.

The two pilots on board had been very experienced. The Captain, Omar Carbore, aged thirty-three years, had a total flying time of 4,576 hours, of which 1,762 hours had been flown on LV-HHB. His co-pilot, Juan Jose Raque Peduzzia, aged forty years, had flown a total of 7,268 hours, of which 1,280 hours were on the accident machine. LV-HHB had flown a total of 25,759 hours, of which 15,848 hours had elapsed since the last overhaul.

The accident was investigated by an Argentinian Accident Board of Inquiry and members of the United Kingdom Accidents Investigation Branch, and a technical member of the manufacturing company flew out at the request of the Argentinian Board, to observe, and to assist, if required. Shortly afterwards, I was in consultaation with metallurgical colleagues at Farnborough who were studying broken pieces of that aeroplane from the Argentinian desert.

Towards the end of June 1976 representatives of the aeroplane manufacturer, the AIB, the Civil Aviation Authority and the RAE gathered together at Farnborough to consider the results of the work so far completed. It became evident at the meeting that certain aspects of the technical investigation remained unresolved, and it seemed that these could best be met by a detailed examination of the remains of the aeroplane out in Argentina. I attended the meeting and pondered over the question as to who would be asked to go, when suddenly and unanimously I found myself chosen to make the examination. I would be accompanied by one of the AIB inspectors and the manufacturer's representative, both of whom had made the earlier visit. It was felt that these two could ease my problems tremendously, leaving me free to carry out my task unrestricted.

We were met at the airport in Buenos Aires by the members of the Argentinian Board of Inquiry and began immediate discussions, explaining in detail the reason for our visit. We then departed for the far side of the airfield to view the wreckage of the airliner at the engineering base of the Argentinian Airline, and learned that only that morning it had been removed to the YPF facility nearby. The National Oil Company had its own aeroplane engineering base at the airport.

At the YPF facilities we found the wreckage heaped together

on the grass, out of doors. Whilst discussing it with the Board and planning its removal into a nearby hangar, for my examination, I was approached by a strange little man who wished to tell me that he knew where a certain part of the aeroplane was. Apparently this part had not featured in the investigation, and, he said, the Board knew nothing of it. He was very mysterious, almost to the stage of being frightened, and talked in very poor English with many hand gestures. I passed him over to the British Representative in Buenos Aires of the aeroplane manufacturer, to discuss the whole matter in his own language to try to authenticate the story. If it was true, the investigation might well be taking on a new twist.

Next morning we drove out to Ezeiza to commence the work on the wreckage. One of the reasons for my 'second look' at the wreckage, at this time, was because, when the fractured end of the starboard wing had been examined, fatigue cracking had been found, and it was this material which had been sent to Farnborough. At that meeting, there had been some difficulty in resolving the wing detachment with the amount of fatigue cracking found at that time. The witnesses had seen the aeroplane apparently moving steadily across the sky, and then the wing had fallen away. Nothing untoward had appeared to have taken place to have initiated that detachment. The design firm had calculated that some additional loading over normal level flight conditions was required to fail the wing. I was now trying to resolve that problem. Was there something else in the wreckage that had been missed – something that perhaps had in flight caused the pilot suddenly to induce that extra loading needed to help the wing? When we heard about the piece of aeroplane from that strange little man, we wondered if it was to prove significant. We had gained the impression that, if this particular piece had become detached, it might have struck the tail unit and this in turn caused the pilot to take some action. At this stage it was all conjecture, but since the subject had been raised, we had to consider it, and there might be evidence in the wreckage to resolve this aspect.

That little man had given no indications as to where the piece of aeroplane was to be seen, and its absence from the wreckage before us prompted us to think that it was out at the accident site in western Argentina. That first morning we were met by the foreman in the hangar. He spoke passable English and had been appointed to liaise with us. He asked if we were interested in the nose-wheel bay port door. The mystery deepened, because that was the piece referred to by the little man. We affirmed that we would very much like to see it, but how could this be arranged? Immediately, as if on cue, a workman walked into the hangar, carrying the door, and handed it over to us. We were also given a brief sketch prepared there and then, as to where it had been found at the accident site. We also learned that the piece had been at Ezeiza for the previous twenty-one days.

We naturally advised the Argentinian Board immediately of this matter; they knew nothing. Because of this new discovery, and the fact that there were still pieces of wreckage at the accident site, the Board made arrangements for the three of us from the UK, together with two members of the Board, to fly out west to Neuquen and travel on to the accident site on Saturday 10 July. The rest of the day was then spent examining the wreckage, and, good to its word, the Board had had all of it transferred indoors and laid out for my examination. We were also given full assistance by YPF personnel.

It was well into Saturday afternoon when we reached the oil rig and well head. We drove straight to the area where the main part of the aeroplane had fallen. There was an amazing and sobering sight before me: the aeroplane had fallen to the desert and had been terribly smashed, and so too had the occupants; the area had been turned into a massive grave, about twenty yards square by $2\frac{1}{2}$ feet high, bounded by wooden crosses bedecked with wreaths as a perpetual shrine. We knew previously that the main wreckage had been interred on site, for we had had serious thoughts of asking for certain parts to be dug up for re-examination. The sight of those crosses and wreaths settled it. In no way could we now ask that the hallowed ground be disturbed.

I then met the man who had actually spotted that nose-wheel bay port door, shown to us at Ezeiza. He described in detail how and where he had found it. I was then given a conducted tour of the whole area so that I could picture the accident scene. By now it was almost dark and we had to return to Plaza Huincul, the nearby oil town, for the night.

We spent the whole of the next day surveying and plotting out the locations of the various pieces of aeroplane that had become detached. The Argentinians had put down markers and although the pieces had been removed weeks before, indentations and ground scars still remained and I was able to make my own wreckage map, pacing out the distances in straight lines, walking straight through scrub and grass-like vegetation, getting scratched in the process. I needed this map to enable me to make trajectory calculations to determine probable heights and orders of detachment of the pieces in the air. The oil company geologist, Pedro Leonardi, accompanied me and with a compass took bearings, enabling me to complete my map.

Monday was spent in Plaza Huincul, consolidating our position, preparing illustrations, holding discussions with YPF pilots, who had suffered a morale loss with the accident to one of their fleet. We also examined a sister ship of the crashed aeroplane that had been grounded since the accident and saw some of the pieces of the crashed machine that had been kept at Cutral Co, the aerodrome for Plaza Huincul. The early hours of Tuesday morning, 13 July, found me sitting next to Pedro as we drove back to Neuquen. We started in the dark, but dawn broke as we travelled across the desert, painting yet another variation on the scenes we had viewed for the past few days.

Tuesday, Wednesday and Thursday were spent back at Ezeiza examining the wreckage and then on Friday morning in Buenos Aires we had a meeting with the Argentinian Accident Board, together with many other specialists who had been co-opted to assist the Board. Saturday was again spent with the wreckage, and Sunday and Monday writing up notes and consolidating material. On Tuesday we left for home.

What had happened to that airliner over the Meseta Buena Esparanza that afternoon in April? And what, if any, was the significance of that piece of aeroplane referred to by the little man who had approached me that first day at Ezeiza Airport?

The ground witnesses at the oil rig had seen parts of the aeroplane fall away. My examinations had been aimed at determining the nature of the detachment of these parts, and also the order in which they had occurred. These examinations were of course complicated by the fact that I had to look at pieces in three different places. I had studied parts of the lower skin from the starboard wing in the Metallurgy Laboratory at Farnborough; then I had seen the remainder of the starboard wing, the majority of the tail unit components and the nose-wheel bay port door at Ezeiza, and the remainder of the tail unit at Cutral Co, Plaza Huincul, in western Argentina. To complete my subjects for examination, I had been given a large number of photographs of the accident site. In the event, I found I had no difficulty in co-ordinating the evidence from the many sources.

My studies of the photographs of the main wreckage enabled me to identify all major components of the aeroplane, with the exception of those that I had been able to examine physically myself. Thus I knew the extent of the structural break-up in the air.

The starboard wing had become detached from the aeroplane in the air and had fallen to the ground about 3,000 feet from the main wreckage. From my examinations at Farnborough and Ezeiza, I could see that the wing had become detached by bending upwards just outboard of the engine position. The separation had commenced in the lower skinning of the wing, and although I could see that a straightforward break had occurred due to that bending process, there was present along the line of separation in the skin a substantial length of failure that must have been present before the accident. That length of failure consisted of fatigue cracking, and in fact extended for about thirty-six inches.

The starboard tailplane and elevator had become detached in the air and were found about 1,600 feet from the main wreckage. I was able to show that these components had failed by bending upwards about the side of the fuselage at the rear of the aeroplane. The upward movement during detachment had carried tailplane and elevator up against the leading edges of the fin and dorsal fin. There was no other damage to the tail surfaces than that associated with the impact with the fin and dorsal fin. Nothing had become detached from the aeroplane and struck the tail to induce the pilot to take any action that may have then caused the wing to be overloaded. And when I examined that nose-wheel bay door, it had not struck anything apart from the ground. The door itself had become detached by being over-opened against its hinges. No weaknesses or pre-separation failures were seen.

I could sum up my efforts as follows: the nose-wheel door had not struck any other part of the aeroplane after its detachment; the starboard tailplane and elevator were not struck before becoming detached; the tail unit items had moved upwards against the fuselage with no evidence of any rearward movement, which indicated that detachment could not have been associated with normal forward flight conditions but was compatible with violent rolling manoeuvres, such as could have followed detachment of the starboard wing.

I had to conclude that the starboard wing was the first item to have become detached. These conclusions were supported by my studies of the accident site, by the disposition of the various pieces and by the trajectory calculations I made of the detached pieces. The problem of why the wing had become detached, when it did, in essentially level flight conditions, with no obvious assistance from anything else on the aeroplane, was now firmly in the court of the manufacturer. However, the studies by the metallurgists had continued whilst I was in Argentina, and I later learned that further evidence of fatigue had been discovered.

It seemed then that the fatigue cracking had made inroads into the strength of that lower skin in the starboard wing, until a stage had been reached where very little excess loading had been needed to fail the wing completely. A check of the rest of the fleet of this type of aeroplane, throughout the world, produced other examples of fatigue cracking in the lower skin. That fatigue cracking had occurred at a location in the wing structure not normally readily accessible to examination, and it was the disruption of the wing in the accident that brought the matter to light. Once it had been discovered, all the other aeroplanes could be examined in a specific manner to reveal their cracks. It was then a matter for the manufacturer to determine the whys and wherefores of the problem so that appropriate measures could be taken to eliminate any possible recurrence.

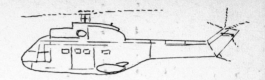

# Ball Bearings Hold the Answer

Great advances have been made in aviation technology, posing new problems for the accident investigator, since I first started in wreckage analysis work in 1941. However, part of the task has remained basically the same – the study and analysis of the wreckage itself. The subject may have altered in character and format because of the technological advances, but success will still largely be the result of the manner of approach and expertise brought to the problem by the investigator.

Advances in wreckage analysis techniques will not always have been dependent upon the advances generally made in aviation, but rather on a particular problem and the flair and ingenuity applied by the investigator in resolving it. I have chosen as my closing example just such a case. It was one of my last major investigations and illustrates the point that, despite the existence and use of commonplace components in helicopters for many years, their part had never before been considered a significant feature to be used to resolve a particular problem. Indeed, back in 1941 such an approach to an investigation would not have been contemplated.

I found this exercise very challenging and exciting too, because in the event new techniques for dealing with some aspects of helicopter accident investigation were evolved and found to be highly successful, thereby providing the modern investigator with another tool.

A Puma helicopter crashed on 28 January 1978 whilst on

exercises and its crew were killed. During the accident flight, part of the helicopter became detached in the air and struck and damaged the tail rotor blades. The helicopter did not crash immediately but flew on for several seconds, apparently normally, and then suddenly pitched nose down as more pieces were seen to break away.

When the wreckage was examined, it was determined that the main rotor blades had become detached in the air shortly before the crash but after they had struck and severed the tail unit. This was a separate event to the tail rotor strike by another piece of the helicopter.

I had already been involved in the investigation but now received a specific request for further study and analysis of the wreckage with a view to determining if any evidence existed that might explain why the main rotor blades had struck the tail boom. This request had been made because loss of directional control, when the tail rotor blades were damaged, should not, of itself, have necessarily led to the helicopter crashing. It was thought more likely that the direct cause of the helicopter falling to the ground would have been the main rotor blade/tail boom strike.

A helicopter is maintained in flight by the rotating blades of its main rotor. These are essentially long, thin wings fastened to a central hub, but they are capable of having their angle changed to increase or decrease the amount of lift that they can produce.

Control mechanisms enable the pilot to alter these angles in flight. If all blades are equally and simultaneously altered, the helicopter can rise or fall, because the overall lifting capacity of the rotor blades has been increased or decreased. This adjustment of angle or pitch of the blades is known as the collective pitch control.

To cause the helicopter to move forwards, backwards or sideways, the pilot can cause the blades around one sector of the rotating rotor disc to increase their angle, and decrease them around other sectors. This moves the lift to one side of the rotor disc and causes the helicopter to move accordingly. Because the blade angles alter as it moves round the disc, this method of control is known as cyclic pitch control.

As each blade rotates round, it is continuously changing its angle and, because of the speed with which it is rotating, tremendous outward forces are exerted on the blade roots, so that large and powerful bearings are necessary to enable the blades to move easily, yet be firmly held.

I felt that the most profitable approach would be to examine the main rotor blade root bearings for the blade that had made the initial strike on the tail boom. The ball indentation damage on the bearing race surfaces, created by the shock loadings arising from the outer end of the blade contacting the boom, could be analysed and a blade pitch angle for the instant of impact determined. This damage could be consequential upon the tip backward blade bending, introduced by the impact. At the same time as this initial impact, I considered that the other three main rotor blades could well have suffered root-bearing damage consistent with forward bending of the blades, due to the momentary deceleration of the rotor disc created by the initial blade strike. I considered that a unique pattern of damage for all the blades' bearings could have resulted, from which I could deduce a set of blade angles, and in turn what the equivalent pilot input on his controls had to be to achieve this.

To help me in all this, I was able to refer to rigging information on main rotor blades for this type of helicopter, so that I could obtain pitch angles all round the rotor disc for high and low collective and for various cyclic stick settings.

First I had to determine which of the four blades had made the initial strike on the tail boom. I rebuilt, jigsaw-puzzle fashion, the bits and pieces of the tail boom and quickly found that the damage made by the blade passed through the blue-coloured roundel along the boom side. Thus I could hope to see metallic scoring and blue paint traces on the blade-leading edge at a specific location from the blade tip.

I examined the four blade-leading edges, after piecing the wreckages of the blades together. Two of the blades were clear of damage at the expected location; one blade was marked with blue paint and scores, and the fourth leading edge was missing. The order of the blades around the disc, when rotating, was first the marked blade, then the 'missing leading edge' blade, followed by the other two. If the missing blade had been the initial striker, I felt it highly unlikely that the two

200

clear blades would have passed through the boom unscathed, and then the marked blade made a strike. I had to conclude that the marked blade was the blade that had first struck into the boom.

I now moved on to examine the main rotor blade root bearings. When such ball bearings are shockloaded, indentations can be made in the surfaces of the inner and outer races by the balls being squashed. By aligning the patterns of indentations, the relative positions of the parts housing the inner and outer races can be determined for the instant of shockloading. The root end of the main rotor blades terminated as a sleeve, and this fitted over a spindle pointing radially out from the main rotor head, or hub. Freedom of movement of sleeve around the spindle (in the blade pitch change sense) was achieved by a pack of bearings.

The outer races of the bearings fitted firmly into the blade root sleeve, and the inner races were tight on the spindle. Thus I could align and relate any ball indentations to pitch angles of the blade. These bearings were about five inches in diameter, with balls of about three-eighths of an inch diameter. Six such bearings made up the pack.

The general location of the aligned damage around the periphery of the bearing would indicate to me the location of the axis of loading across the bearing to produce such damage. Thus in planform, where the blade was subjected to a tip backward bending loading, ball indentations might be expected around the trailing edge of the bearing in the pack nearest to the hub, and around the leading edge of the outermost bearing in the pack. Because all four main rotor blades had become detached in the air, by failing near their roots, there was little likelihood, I argued, of ground impact forces producing any damage patterns on the root end bearings. Any evidence of ball indentation would therefore be relevant to the event I was considering.

I started with the bearings of the blade that had made that initial strike on the boom, marking them so that I could relate any damage areas to vertical and horizontal datum lines. I quickly discovered ball indentations present on the innermost bearing, and I was able, under the microscope, to study individual damage areas for each ball and to plot them onto a

201

drawing. I did this for the inner and outer race tracks. Interestingly, the indentations were located around the upper rear portion of the bearing (upper part of the trailing edge), which indicated a tip upward and backward bending loading through the blade.

I realized that the blade would have had to make a steep descent around the rear starboard quarter of the rotor disc to contact the tail boom, so the upward component of loading on the bearing was not unexpected. The nature of the loading on the bearing confirmed that this blade had indeed struck the tail boom. I then aligned the damage patterns of the inner and outer races, and this showed that the blade must have been at a geometric pitch angle of approximately +2° at the instant of impact.

I now turned to the blade with the missing leading edge, and next in order of rotation around the rotor disc. This examination showed that ball indentations were present on the innermost bearing and again, as on the previous blade, were clearly defined. This time, however, the evidence was located around the forward (leading edge) section of the bearing, and thus indicated a tip forward bending loading through the blade. I aligned the damage patterns on the inner and outer races and this showed that the blade must have been at a geometric pitch angle of approximately $-2\frac{1}{2}°$ at the instant of impact.

The next blade in the disc contained ball indentations on the innermost bearing, and these were around the forward section, indicating tip forward bending loading. The geometric pitch angle this time was determined as +12°.

The final blade's innermost bearing contained ball damage, again indicating a tip forward bending loading, and when I aligned the ball damage, the geometric pitch angle came out at approximately +10°.

I had considered that a unique pattern of damage could result from one blade striking the tail boom, and my examinations of the blade bearings produced evidence of such a pattern. Thus, when the one blade struck the tail boom, it received tip backward (and upward) bending loading and the other three blades did swing forward and receive appropriate loading at their root bearings.

Thinking about the geometric pitch angles – when the pilot moves the cyclic pitch control in a particular direction – this action will be seen at the blades at ninety degrees to the line of application; for example, a fore and aft stick application affects the angles of the blades in the beam positions in the rotor disc, and lateral stick application affects the angles of the blades in the fore and aft positions.

I now knew all the blade positions and angles for the instant of tail boom strike and, from the foregoing, could interpret these in terms of cyclic pitch control column positions. I referred to the rigging information and found that my blade angles related to the cyclic pitch control column having been fully to one side, and half back from centre, all with a low collective pitch setting. It required a specific combination of collective and cyclic pitches to achieve the pattern of pitch angles I have found from my examinations. It was highly unlikely that this could have been attained by a fault or failure in the helicopter control system, but could be obtained by discrete control inputs made by the pilot.

Thinking about the events immediately prior to the main rotor blade/tail boom impact suggested to me a likely sequence of control inputs which could readily result in the angles I had determined from the blade root bearing damage.

As a result of the preceding damage to the tail rotor, the pilot could have made instinctive reactions to the consequent helicopter movements, and to prompt him to make an entry into auto-rotative flight. This would have been achieved by lowering the collective pitch and applying lateral and aft cyclic stick. Thus the natural consequence of the tail rotor blades being damaged would be the production of the main rotor disc pitch configuration that I had found. However, steadily applied cyclic stick inputs could not have placed the rearward blade sufficiently low to strike the tail boom, and the attainment of such a low attitude by the striking blade indicated to me that there must have been a sudden high-rate cyclic pitch stick movement coupled with the introduction of minimum collective pitch.

The standard operating procedures taught to single-engine and low-inertia rotor disc helicopter pilots is to lower fully the collective pitch before entering auto-rotative flight. On the subject helicopter, however, which had two engines and a high-inertia rotor disc, it would be necessary only to lower collective pitch to approximately +8°. It seemed to me that in the emergency that arose on the accident flight, whilst the appropriate corrective sense control inputs were made, that input for collective pitch had been made to excess, and that for cyclic pitch had been made at far too high a rate of application.

The pilot in question had received all his earlier and formative training on single-engine and low-inertia disc machines, and it did seem likely to me that here was a possible case of a man 'reverting to type' in an emergency. The pilot had suddenly been confronted with a serious problem of a known nature and had reacted instinctively in the manner of his early training.

This was all very interesting because we then learned that two other helicopters of the same type had suffered similar tail boom strikes by main rotor blades during enforced landings, overseas, as a result of the pilots making premature flares from auto-rotative flight – in other words, over-controlling rapidly.

I had just completed nearly forty years of wreckage analysis work with a very satisfactory conclusion to an intriguing problem, but then to me all of the hundreds of problems over the years had been equally intriguing, deserving maximum attention. All too often, wreckage, being an untidy-looking and meaningless mass of debris, can been pushed away into an obscure corner not conducive to careful detail work. The lesson it has to teach may then be lost. We should remember that 'A wise man learns from the folly of others, a fool seldom by his own', and of course the learning in this case can lead to the saving of many lives and aeroplanes.

There are of course some events that bring about aeroplane accidents where the discovery of the cause cannot of itself necessarily lead to the raising of safety standards because they simply highlight the fact that most problems arise from man himself and not from the machines he has designed and created. In the next Chapter I refer to such events which show that the sudden and horrific increase in aviation fatalities in mid 1985 did not indicate a lowering or decline in safety standards.

# CHAPTER 23

# *Epilogue*

Safety in the air has been improving steadily for many years and in 1984 only 101 fatalities occurred on scheduled airline flights. In 1983 the figure would have been about 700, but was increased to 1,200 by the loss of a Korean airliner shot down over Russian airspace, and other acts of sabotage throughout the world.

The annual passenger fatalities total for each of the years 1964–83 would be about 750. However, the best single measure of aviation safety is the number of fatal accidents per flight or rather million flights. The average passenger is more interested in numbers that reflect his prospects of completing his flight, rather than an obscure statistic on some annual basis.

About fifty years ago the British Accident Rate was 0.67 fatal accidents per 10,000 flights, that is, 67 for every million flights. With passenger flying still very new and limited to the few who could afford to fly, accidents were few and far between. Today, the World Accident Rate is below 2.0 fatal accidents per million miles, and the number of flights made in the period 1973–84 was over one hundred million. To use the number of fatalities as a yardstick belies the safety standard of aviation today, because airliners are carrying many more passengers on each flight than ever before – often over 500. Consequently, a single accident can almost account for a year's fatalities in former years.

This has been so in 1985, where two 'jumbo jets' crashed

with a total death toll of 849, and a smaller airliner suffered an engine failure and fire on take-off resulting in 54 deaths. Thus three accidents alone accounted for over 900 fatalities.

This of course is a horrific situation in terms of human loss, but in aviation terms, quite a different picture emerges. Aeroplanes are really no less safe than in 1984, and the apparent decline in standards is quite simply due to the frailty of man and not of machine. Errors and actions of human beings, for example lack of good engineering practice, common sense and logic in diagnosing problems that arose in service, accounted for 594 deaths, and irresponsible action (i.e. terrorists) for another 329.

Before considering the actual cases, here is an example of the type of thinking that leads to such failings. First a non-aviation situation, a fairly common occurrence is a household: an electrical appliance ceases to work; the fuse is examined and found to have failed. Many people will change the fuse – repeatedly if necessary – if the appliance still fails to function, and the fuses will continue to fail. They will not check the appliance itself for the reason of failure. In other words, they apply a remedy to the symptom and not to the cause. In consequence, the real culprit remains undetected and the repeated application of electrical energy may eventually destroy the appliance.

Surely such a situation or approach to a problem would never arise in aviation; but it has – on more than one occasion, and recently too, and with horrific consequences. But first, let me take an example of an earlier case of symptoms being treated, rather than causes.

In 1956, one of the V-bombers was being used at Farnborough in a variety of roles. It was instrumented and equipped for measuring such things as structural vibrations and deflections and other parameters of interest to aerodynamicists and structural engineers. One such experiment required the use of electrical energy, and conveniently this was drawn from one of the bomber's domestic services at the navigator's station.

It must be pointed out that this particular type of aeroplane was rather unique: its controls were fully electrically operated and not by some form of hydraulic power assistance as with other contemporary aeroplanes of such a large size.

As a type, this bomber had been flying quite successfully and had never experienced loss of control due to major electrical failure. As far as possible, the designer incorporated safeguards against all possible eventualities. The safeguards included not only the usual fuses – much like those used domestically in the home, only larger – but also circuit breakers. These are springloaded switches which 'pop out' under excessive elctrical loading, thereby protecting the associated circuit and equipment.

On Friday 11 May 1956, just after lunch, the bomber left Farnborough to carry out a flight, primarily for a scientist to conduct the experiment connected to the electrical outlet at the navigator's station. The aeroplane had full fuel tanks and was likely to be airborne for several hours. On the way to the south coast, the scientist switched on the experiment. Almost immediately the appropriate circuit breaker 'popped out'. Imprudently, the scientist reset it, and it again 'popped out'. The scientist reported the situation to the pilot that his experiment could not be carried out, and because this was the main purpose of the flight, the pilot decided to abort the flight. However, before returning to Farnborough he had to use up a large quantity of fuel to bring the landing weight to an acceptable figure. He had taken off fully laden and was now returning very much earlier than originally intended.

The pilot set up a low level flight pattern over the sea off the south coast with conditions set for maximum fuel consumption. The flight continued quite satisfactorily for some considerable time. During this, the bomber was flying in and out of low cloud near to Shoreham and Brighton.

Now it seems that while this was being done, the scientist could not leave well alone, and in went the circuit breaker, not once but several times. Suddenly the pilot felt a stiffening of the controls and saw indications on the instrument panel of an electrical failure of major proportions. He immediately eased the bomber into a climb to make height for he now had a major emergency on his hands.

The bomber emerged from cloud at about 1,000 feet over Shoreham. Due to the electrical failure the flying controls were completely unresponsive. The pilot ordered everybody to eject.

They were now descending fast towards Shoreham railway station. At almost the last second, one ejector seat was seen to leave and its occupant made a short but safe descent, landing not more than fifty yards from where the bomber struck the railway embankment, near the station. All other crew members perished in the crash although the escape hatch was found away from the bomber. The aircraft disintegrated upon impact and debris was strewn for hundreds of yards across the playing fields of a nearby school. Mercifully, the fields were clear of pupils at the time.

Within an hour I was being flown down to Shoreham from Farnborough, and as it was only twenty minutes flying time, I was to make this trip several times in the next few days. Discussing the accident with the manufacturer's design staff as we walked around the wreckage, some of the story of the flight began to emerge. A certain amount of information had already filtered through from Farnborough Control Tower and of course there was the surviving co-pilot. It was clear that there had been a major electrical failure and in company with the manufacturer's chief electrical engineer, we paid special attention to the state of any electrical components that we could see.

The engineer appeared next morning having spent much of the night working on a hypothesis to account for much of what we knew. Could the wreckage evidence support his theory, or would it produce some other explanation?

We isolated as much of the electrical material as possible from the wreckage and this was sent to Farnborough. The remainder of the wreckage was to follow. There was an air of supreme urgency as other aeroplanes of the type could have been at risk. Our searching and examinations, along with the work of the electrical design staff at the manufacturer's works, had shortened the list dramatically. It now transpired that if, as we suspected, the circuit breaker had been repeatedly reset, then the consequences must have included the 'blowing' of many fuses, but one in particular – the navigator's station equipment circuit – would really prove the theory.

My priority task became the search for that one fuse to determine if it had 'blown' electrically. The fuse in question was about three quarters of an inch in diameter, and about one and a quarter inches long. It consisted of a heavy ceramic body, with metal end caps, attached by bolts to heavier items known as bus bars. There were many such fuses fitted in banks or rows

208

in the bomber and although originally marked with paint for identification purposes they were broken and now devoid of such identity. They were also mixed with tons of twisted and burnt looms of wire and metal.

My fuse, either intact or in pieces was in the large mound of wreckage now at Farnborough. It was the Whit Monday holiday and I planned an assault on that mound. I had the help of skilled, semi-skilled and unskilled people. The approach was simple. I formed a pyramid of searchers with the unskilled at the mound. They were briefed to sort out all wreckage of a certain character. I made sure they would include more material than I wanted to ensure that the right stuff would come through for searching. Unwanted wreckage was discarded and required material passed back to a second row of helpers who had been briefed in more detail. They in turn passed back this selected material, and so the search continued. Each row of searchers was more qualified to select than the one before, and I was at the apex of the pyramid to make the final decision.

By patience, persistence and careful searching and study, we discovered the first of the end caps after only two hours. It was still bolted to a piece of identifiable bus bar, although this itself had been badly damaged. An hour later the other end cap was found. Both caps contained pieces of ceramic body – protruding but not mating to make up the body. We found other odd pieces of ceramic and one of these made the link between the end caps: I had my fuse.

I took the end caps to my microscope and there was the evidence that I sought. Traces of fuse wire inside each cap bore witness to the fact that the fuse had been electrically overloaded. The wire ends had formed into globules – a tell-tale sign of a fuse melted by electrical overheating and not by crash fire burning.

We now knew that the electrical supplies in the bomber had been disrupted by the repeated resetting of the circuit breaker – a senseless, illogical action which had led to tragedy and loss of life, including the life of the instigator of the situation. The repeated resetting of the circuit breaker was akin to tackling the symptom and not the cause of the experiment failure.

However, the accident prompted a long hard look at the electrical circuitry of all the large bombers and some shortcomings were brought to light. As I have said on other occasions, regretfully, in aviation someone must suffer, it seems, if aeroplanes are to be made safer.

Twenty-nine years later, a similar situation arose whereby a symptom was dealt with rather than the root cause of the problem, and this led to a horrific event at Manchester Airport in August 1985 with the loss of over fifty lives. All of this despite the fact that at one stage in the proceedings, an engine manufacturer had suggested a cause for certain symptoms which had been a source of concern. After the accident, investigators found that that very cause had been responsible for this terrible tragedy.

The aeroplane involved was a Boeing 737 twin-engined airliner. It was powered by two turbo jets. In simple terms, the engines consisted of an air intake at the front, leading to two compressors – one high pressure and one low. Next to the compressors was the combustion chamber, where the air was mixed with an appropriate fuel, ignited and burned to then pass aft through low and high pressure turbines, the rotation of which then propelled the hot gases along a tailpipe.

The combustion section comprised nine cylindrical flame tubes, lying lengthwise around the engine and equally spaced just inside the engine casing. The flame tubes included a burner at the forward end inside a domelike nose, and a discharge orifice at the rear which directed the hot gases into the turbine section. When fitted to the aeroplane the engines were positioned under the wings with the No. 9 flame tube at the top. It was separated from an integral fuel tank inside the wing by the thick wing skin, the engine cowling and engine casing.

The Boeing 737 airliner, as a type, was twenty years old and had never had any major problems. The engines were subjected to routine and regular inspections and maintenance. In the case of the engines on the 737, the flame tubes were to be given a first inspection after 3,300 take-offs, and in the case of the subject aeroplane, this took place in September 1983. The inspection revealed a crack about seven inches long in a flame tube, at its forward end on No. 1 engine (port). This crack was repaired by welding by the airline operator. At the time, the engine manufacturer had issued a warning to all operators that a welding repair was not as strong as the original metal and

210

recommended that either a solution heat treatment or stress relieving treatment be carried out to such weld repairs. The operator did neither and the Civil Aviation Authority approved the repair as done and the aeroplane continued to fly in service on holiday traffic flights to the continent.

On a number of occasions the engine was found to be slow in coming up to power for take-off. The idling revolutions were also low. The operator looked into the problem and decided to resolve it by adjustments to the engine bleed valve and by changes to the fuel controller unit. The engine manufacturer recommended to operators of this type of engine that they 're-affirm their procedures' for investigating slow starting or acceleration characteristics: 'Liner cracking and flame tube misalignment in some cases is evidenced by slow run up from stop to idle conditions, or by slow acceleration above idle.' The manufacturer advised that if the flight crew noticed these characteristics, they should proceed according to an appropriate section of the maintenance manual. Failing this, the engine combustion section must be inspected before the next flight.

The subject aeroplane and engine had not been subjected to this action and continued in service on the busy holiday traffic. On Monday 19 August 1985, the aeroplane took off for England from Crete. It landed three and a quarter hours later and was refuelled for another flight. One hour later when being run up for take-off, the No. 1 engine was slow to come up to power and needed more throttle application to balance with the No. 2 engine. In fact, two inches of extra movement of the throttle lever was necessary. The fuel controller unit was changed and the problem 'seemed to be partially solved'.

At 2.15 p.m. on 21 August 1985, the aeroplane arrived at Manchester from Barcelona. The crew reported that the No. 1 engine had slow acceleration times and an abnormally low idling speed. The operator effected some adjustments which seemed to balance the engines again and the aeroplane continued in service. A flight of three and a half hours to Athens was uneventful, and after a turn around, the aeroplane returned to Manchester and landed.

The next morning, at 7.00 a.m. with 131 passengers on board for a flight to Corfu, the aeroplane was prepared for take-off. At

7.12 a.m. the engines were run up and the aeroplane moved along the runway. Thirty-two seconds later a loud bang was heard, both on the aeroplane and all over the aerodrome, and the aeroplane was seen to slow down and turn off the runway. A fire was seen which became very intense and evacuation of the aeroplane was ordered. For various reasons this was not wholly successful and fifty-four passengers perished in the aeroplane. The aeroplane had not even left the ground when the accident occurred.

It became readily apparent from the investigation that followed that the No. 1 engine had suffered a failure – the result of which was a catastrophic fire. Detaching pieces of engine had cut into the wing fuel tank above the engine to release fuel which then poured down on to the exposed hot parts of the engine. The nose portion of the No. 9 flame tube was found on the runway near the scene.

Examination of the flame tube showed that the nose section had become detached because of a long crack that had been present around the tube before the accident. Detachment had been inevitable.

My analogy to the setting of the circuit breaker in that earlier accident now becomes apparent. The operator had persisted in following up an engine problem by adjustment and replacement of fuel system items, although it was not effecting a cure, and despite the fact that the engine manufacturer had advised that such symptoms indicated a cracked flame tube as a cause, and recommended an inspection before the next flight, nothing more was done. This is an accident which surely need not have happened and wrongly indicates that aeroplanes are unsafe.

This particular accident also highlighted the use of inflammable materials for the furnishing and upholstery of aeroplane cabins, and questions were being asked why the use of such material continued when technology had now produced suitable non-flammable materials. The answer shows the continuing policy among airline operators of putting profits before passenger safety.

One of the most critical aspects of aeroplane design is weight. The end product is always a compromise because the designer has to balance three main elements of weight. First there is the actual structural weight of the aeroplane, its controls and engines. This all has to be of sufficient size and strength to perform the role required and therefore must have an appropriate weight for the materials used. This weight cannot alter once the design of the aeroplane has been fixed. The second element of weight for consideration is the payload or usable load, such as passengers or cargo in a civil aeroplane. The third element is fuel. A certain amount of fuel must always be kept for emergencies or diversions in the event of weather problems at the intended destination, but the aeroplane may not necessarily carry a full fuel load for every flight. Thus, of the three weight elements making up the total weight for a flight, one (the aeroplane itself) is a fixed quantity and the other two are variable. To achieve maximum range may mean limiting payload, etc.

In commercial flying, flights are made for profit, and this means carrying maximum passengers or cargo on every flight. Every empty seat means less revenue. If extra weight is required to be added to the aeroplane itself, this in turn means a loss of weight elsewhere. In commercial flying, this usually means loss of passenger seating. This aspect becomes an important factor where safety is concerned, particularly in the modern passenger machine where much attention is given to passenger appeal and the right furnishings and upholstery must be chosen. Unfortunately, man-made fibres are not usually fireproof and if ignited can give off noxious and deadly fumes. This is a very valid point in the event of a fire in a crash, as was the case in the accident at Manchester, where passengers were overcome by fumes and flames preventing successful escape from an otherwise intact aeroplane.

In the clamour for safer materials, a problem is then posed to the designer and the operator. The scientist and engineer may well produce a suitable material but find that it is slightly heavier than that in current use. For example, in a modern 'jumbo jet' cabin there maybe about 350 seats. The replacement material for each seat might add say three pounds. This means 1,050 pounds extra weight on the aeroplane. If the average passenger is 140 pounds in weight this extra weight is the equivalent of about seven passengers. This could mean the removal of seven seats, in which case the operator will seek to recover the loss of revenue by increasing seat fares. Such an action may make the flight less attractive to a prospective passenger and so more than the original seven seats may in fact be lost. The operator is then posed with the choice: greater safety in the event of an accident or incident involving fire, and loss of revenue; or lower safety and same profits. With aviation becoming ever safer, the likelihood of the fire hazard arising lessens, so there must be a tendency for the operator to go for the option of keeping the old seat and his profits, until of course the scientists come up with a lightweight material but with safer characteristics.

Turning now to the two 'jumbo jets' that crashed with 849 fatalities. Was this the result of lessening standards of safety? I think not. In the first case the aeroplane had been converted to carry over 500 passengers, against 350 originally. This was done to use the aeroplane on 'short haul' flights – one hour instead of eight or nine. The weight of the extra passengers was offset by the need to take less fuel.

The aeroplane in question was being operated by Japanese Airlines, and on 12 August 1985, it took off on a short flight of only fifty-four minutes. It was carrying 520 passengers and crew. About thirteen minutes into the flight, the captain radioed the control tower at Tokyo, from where the flight had departed, to report an emergency and to request a change of height and of heading to return to Tokyo. Some two minutes later the control tower asked the nature of the emergency. There was no reply. One minute later the aeroplane was again contacted and this time the message was 'now uncontrollable'.

The aeroplane continued flying over a somewhat circuitous route for a further thirty minutes until it crashed into mountain country at an altitude of about 9,000–10,000 feet. The aeroplane was completely destroyed and all aboard except three were killed instantly. The survivors included one of the stewardesses whose testimony was extremely useful.

In the hours and days that followed, parts of the tail unit of the aeroplane were recovered from the sea, at the position where the emergency had been declared. A flight data recorder and cockpit voice recorder were recovered from the wreckage and the traces analysed. Together with the report from the stewardess, and a study of the wreckage an intriguing story emerged. The rear pressure bulkhead, that is, the dome-shaped

214

bulkhead sealing the rear of the pressurized fuselage, had failed in flight, and the escaping air had penetrated the tail unit, literally blowing away the fin and rudder. Unfortunately the detachment of these items also disrupted the tail unit flying controls and the hydraulic power supplies to the aeroplane itself. By a remarkable feat of flying, the captain kept the aeroplane in the air, with no fin and rudder and no means of controlling the machine adrodynamically. Only judicious use of the engines enabled him to maintain some element of control. Regrettably, after thirty minutes all was lost and the aeroplane crashed.

The detailed study of the wreckage showed that the rear pressure bulkhead had failed initially along the line of a horizontal joint extending from the centre to the extreme edge on the starboard side of this very large circular structure (approximately 15 feet in diameter). We must now go back to 1978. The crashed aeroplane had suffered damage following a heavy tail-down landing, scraping the tail end along the runway.

The bulkhead consisted of 18 segments of metal sheeting overlapping along radial lines radiating from the centre of the bulkhead. In the original construction the segment sheets overlapped sufficiently to be joined by two parallel rows of rivets. A stiffener member was also attached along one of the rows of rivets. If a repair was necessary, this was simply done by including a long single strip of doubler plate along the central length of the joint and sandwiched between the segment sheets. The same two rows of rivets were used and the stiffener member; additionally a third row of rivets secured the doubler plate to one of the segment sheets. Examination of the pressure bulkhead after the crash in August 1985, showed that the failure had in fact taken place along the repaired joint from the 1978 incident. The repair had been carried out by a team of engineers from the aeroplane manufacturer, but the examination now showed that it did not conform to the repair outlined above. Doubler plating had certainly been included between the segment sheeting, but not as a single strip. Two parallel strips had been used, laid side by side with a small gap between them. Unfortunately, the gap was so positioned that it was bridged across the joint's reinforcement by only the single thickness of a segment sheet. This of course was counter to the

215

intention of the original repair and in fact constituted bad engineering practice. It is really not surprising that a fracture developed along the segment sheeting alongside the gap in the doubler plating. And so the action of man led to the loss of the aeroplane and 520 lives, again not an indictment of aviation safety standards, but of human carelessness and irresponsibility.

The remaining 329 victims from the three accidents I referred to earlier were all lost when an Air India 'jumbo jet' plunged into the Atlantic off the southern Irish coast during a flight from Canada to England. On 23 June 1985, an Air India Boeing 747 left Toronto with 329 people on board. It was bound for Bombay, but first calling at Montreal and London. All was well until at 7.13 a.m. when, flying at 31,000 feet altitude, off the southern coast of Ireland, its symbol suddenly disappeared from the radar screens at Shannon. Search of the sea under the last known flight position produced bodies and debris from the aeroplane. At the area where radar control was lost, aeroplanes are controlled only by secondary surveillance radar. This means that responses from a transponder on the aeroplane appear as symbols on the radar screen. The picture included aeroplane identification and flight level reading. In the case of the Air India aeroplane, the radar picture had disappeared in a manner indicative of transponder failure or loss of power to it. At the time of disappearance the flight level reading had been steady at 31,000 feet.

The surface search recovered less than half of the aeroplane occupants and most of these apparently came from the rear of the fuselage. Subsequent search and salvage work located and recovered about 20,000 pounds weight of wreckage. This was a remarkable achievement because the wreckage was lying over an area 7 miles by 2½ miles at a depth of 6,700 feet, and 160 miles from the nearest port.

The voice and flight data recorders were recovered from the wreckage and together with the recovered wreckage were despatched to India for investigation. The sudden loss of radar response with no apparent height loss suggested that the aeroplane had suffered a sudden and catastrophic event whilst in level flight. No adverse reports had been received from the aeroplane during the Atlantic crossing and experience of accident matters and knowledge of the type suggested to me that this was possibly the outcome of some extraneous event and not due to failure of airframe, engines controls etc. The recorders were played back and provided no specific clues other than that everything had suddenly stopped without warning. A bomb was a likely suspect. Indeed a telephone call was received in Canada from a man claiming to have planted explosives on the aeroplane. Coincidentally, on the same day that the Air India aeroplane crashed, a flight from Vancouver landed at Tokyo's Narita airport and two baggage handlers were killed by a bomb in the luggage they were handling from the aeroplane.

It was reported that the aeroplane had arrived in Tokyo about ten minutes ahead of schedule. If it had been on time it is likely that the bomb would have exploded as the aeroplane was approaching the airport. Interestingly, the Air India jet was running twenty minutes late, and had it been on time, it would have been approaching London's Heathrow airport at about the time it disappeared from the radar screen off Ireland. Since the accident, two men have been arrested and charged in Canada with offences concerning explosives and the aeroplane.

As a result of the Indian investigations, it came to light towards the end of November 1985, that evidence had been found in the wreckage to indicate an explosion in the forward cargo hold of the aeroplane. I was very interested in this accident and had read all available material relating to it as well as discussing it with persons concerned with the investigation. I had concluded that in the absence of any positive evidence to the contrary only a bomb or explosive device could have accounted for the accident in those circumstances.

And so in the short period between 23 June 1985 and 22 August 1985 903 people died in air crashes, not because of the lowering standards of aviation safety, but because of the frailty or wickedness of man.

# Glossary

# Glossary

**Aerofoil** A suitably shaped structure that generates lift when moved through air, i.e. wing, tailplane, rotor blade.

**Aileron** A movable control surface hinged to the trailing edge of each wing (controls rolling).

**Airbrakes** Surface, usually on wing or tail, that hinges outwards to act as brake to slow down the aeroplane.

**Altimeter** Instrument that registers height above sea level or aerodrome.

**Artificial Horizon** A gyro-stabilized flight instrument that shows pitching and rolling movements of the aeroplane.

**Asymmetry** One-sided, i.e. one engine only working on twin-engine aeroplane.

**Auto-pilot Flight System** Gyroscopically controlled device that keeps the aeroplane in steady flight or puts it through pre-set manoeuvres or patterns as selected by the pilot.

**Automatic Observer Panel** Instrument panel (like pilot's) but with cine-camera recording the readings for subsequent analysis. Early version of flight recorder.

**Automatic (Geared) Tab** Tab hinged to trailing edge of control surface and linked to the main surface ahead so that tab moves in opposite sense to control surface, thereby assisting its movement.

**Backtrack** Taxying along runway against the direction of take-off or landing.

**Banking** Tilting of wings as the aeroplane turns.

**Bomb Slip** Mechanism to suspend a bomb in an aeroplane, and release it when required.

**Boost Pressure** For internal combustion engines: to compensate for reduced atmospheric pressure with altitude, a supercharger increases or *boosts* the delivery pressure of the fuel/air mixture to the engine.

**Bunt** Sudden nosing down of aeroplane in flight, to very marked degree.

**Centre of Gravity** Point at which the total weight of the aeroplane is considered to act.

**Chord** The distance from the leading edge to the trailing edge of a wing, tailplane or fin.

**Cockpit Canopy** Transparent covering over pilot's cockpit on small aeroplanes.

**Collective Pitch Control** Control that varies the blade angle of all helicopter blades simultaneously, to make the craft rise or sink.

**Control Surface** A movable surface, e.g. aileron, to control the flight of an aeroplane.

**Cowling** Covering around the outside of an engine to streamline or smooth the lines of the aeroplane.

**Cyclic Pitch Control** Control that varies helicopter blade angles as they rotate to cause the craft to change direction of movement.

**Delta** An aeroplane with a triangular or delta-shaped wing plan with the point at the front.

**Dorsal Fin** The vertical surface blending the top of the aeroplane body with the leading edge of the bottom of the fin.

**Download** Loading which pushes wing or tailplane downwards from above, bending it about its fixed end.

**Drag** the air's resistance to moving objects.

**Elevator** A horizontal control surface on the trailing edge of the tailplane (controls climb and descent).

**Elevator Mass Balance Weight** Heavy weight to counter-balance airloads on control surface.

**Elevon** Control surface on trailing edge of a delta aeroplane acting as both elevator and aileron.

**Engine Stall** Breakdown of normal airflow through the compressor of a jet engine.

**Fatigue Failure** Progressive cracking of a metal item, usually under fairly normal loading conditions. Leads to aeroplane breakage without resorting to excessive loadings and manoeuvring.

**Fin (Vertical Stabilizer)** The fixed vertical surface at an aeroplane's tail which helps to control roll and yaw.

**Flap** An adjustable hinged surface along the trailing edge of a wing that increases lift or both lift and drag. Used for take-off and landing.

**Flight Director** A flight-deck instrument that tells the pilot whether he should guide the aeroplane left or right, up or down, or stay level and on a present heading.

**Flutter** An unstable air induced oscillation of an aerofoil – an uncontrollable shake.

**'G' Force** Force experienced in a turn or pull up from a dive. Crew feel pressed downward and the aeroplane can be excessively loaded.

**IFR** Instrument flight rules, procedures to be followed when a pilot is flying on instruments without any outside references or cues as to his attitude or position.

**ILS** Instrument landing system which guides a pilot during landing by means of two sets of radio beams transmitted from the ground near to the runway.

**Intake Grid** Mesh covering the air intake to an engine to prevent the ingress of foreign objects.

**Jet Engine** An engine which propels an aeroplane by means of the reactions of its exhaust gases.

**Knot** A speed of one nautical mile per hour, equal to 1.15 miles per hour.

**Leading Edge** The front edge of a wing, tailplane or fin.

**Longitudinal Oscillation** an up-and-down motion of the aeroplane nose during forward flight.

**Mach Number** The ratio of the true airspeed to the local speed of sound. The latter varies with altitudes so a given Mach number does not represent a fixed speed.

**Machmeter** Instrument which presents the speed as a Mach number.

**Magazine** Bay or compartment housing ammunition.

**Nacelle** The housing on a wing to contain an engine.

**Pitch** The angular setting of a propeller blade or a helicopter blade.

**Pitching** Up or down movements of an aeroplane nose.

**Pitot Head** Forward-facing open-ended tube measuring the dynamic air pressure as an aeroplane moves forward (increases with speed). It is surrounded by an outer tube with side perforations to measure 'static' atmospheric pressure. Collectively the pitot provides pressure information which is translated as speed and altitude on the air speed indicator and the altimeter.

**Radome** Non-metallic covering over the radar aerials on aeroplane, usually at nose and shaped to fit lines of aeroplane body.

**R/T** Radio/Telephone – a means of communication between aeroplane and ground by direct speech.

**Roll** Rotation of an aeroplane about its nose-to-tail line.

**Rolling (Pull Out)** The recovery from a dive where the nose is raised and the aeroplane is simultaneously rolled to change direction.

**Rotate** To pull up nose of aeroplane on the take-off run at a speed which permits the aeroplane to lift off the ground and fly.

**Rotor** An assembly of moving wings and their hub, usually rotating in a horizontal plane as on a helicopter.

**Rudder** A vertical control surface hinged to the trailing edge of the fin (controls direction).

**Servo Motor** A motor control which receives small inputs and delivers large outputs – used on an aeroplane control system to magnify the pilot's effort to a control surface, overcoming the air resistance trying to prevent its movement at high speed.

**Slipstream** Airflow produced by a propeller.

**Span** The distance from wingtip to wingtip across the aeroplane.

**Stall** The loss of lift due to excessive angle of wing to airstream or to insufficient speed of aeroplane. The aeroplane can fall away out of control.

**STOL** 'Short Take Off and Landing' aeroplane for operating into and out of restricted areas.

**Tab** small movable surface fixed to trailing edge of main control.

**Tailplane (Horizontal Stabilizer)** The horizontal aerofoil at the rear of the aeroplane.

**Trailing Edge** The rear edge of a wing, tailplane or fin.

**Trim** To adjust the tabs or other control surfaces so that the aeroplane can fly at the correct attitude without manual effort.

**Turbine Shroud Ring** The fixed ring of deflecting blades against the rotating blades of the turbine to straighten air flow.

**Undercarriage** The leg and wheels assembly supporting the aeroplane on the ground.

**Upload** Loading which pushes wing or tailplane upwards from below, bending it about its fixed end.

**Variable Incidence Tailplane** Tailplane that can have its angle altered by the pilot in flight.

**VFR** 'Visual Flight Rules' – procedures to be followed when pilot is flying visually with outside references.

**VTOL** 'Vertical Take-off and Landing' aeroplane with jet engines capable of lifting it vertically for take-off and lowering it vertically for landing – without forward movement in either case.

**Yawing** Movement of an aeroplane's nose to left or right.

**Wing Loading** The weight of the aeroplane divided by its wing area, can give a measure for comparison with other aeroplanes.

# Appendix

APPENDIX

# Aeroplane Accident Investigation in the United Kingdom

In 1913, when the Aeronautical Inspection Directorate was set up at Farnborough, among its responsibilities was the investigation of aeroplane accidents. After the First World War, when two Air Transport Companies began to operate between London and the Continent, it was clear that the safety of civilian, fare-paying passengers had to be guarded also. The Accident Investigation Branch (AIB) was then originated with the Air Ministry.

Those responsible for its establishment acted wisely in two respects. First the Branch was manned by civilians so that sound or well-trained men would not be lost through positions or promotion, and second, the Inspector-in-Charge was made responsible to the Minister of the Department of the day – in this first instance the Air Minister himself – so that he would be able to speak without fear or favour.

It was also agreed by domestic arrangement that the services of the AIB would be made available to the newly created Royal Air Force, if required. The Branch was established in 1920 under its first Chief Inspector, Major J.P. Cooper MC, its prime task being to investigate accidents and to determine the cause, thereby preventing recurrence of the event. The object was not to apportion blame with a possible view to disciplinary action or prosecution. This is still applicable today.

In those early days, with only a small staff, each inspector dealt with all aspects of an investigation, but since World War II the Branch has expanded and two types of inspector have been employed. One has dealt with the operational and flying aspects of

227

an investigation, and the other with the engineering and technical matters, including the examination of the aeroplane wreckage.

It is the engineers today who assist the RAF Boards of Inquiry if so required. For small civil aeroplane accidents, an Operations Inspector and an Engineering Inspector work as a team on the investigation, but for large civil aeroplane accidents a Senior Operations Inspector will be in overall charge with several inspectors of both disciplines dealing with the many and varied aspects of the investigation. The Chief Inspector renders his report to the appropriate Minister, and this is then published and made available to the general public.

Where a major disaster occurs, with fatalities, the Minister will call for a Public Inquiry, under an independent Commissioner, usually an eminent member of the legal profession, supported again by eminent members of the flying and engineering professions as assessors. The Inquiry does not investigate the accident as such but considers all the evidence which is obtained by the AIB and others during the normal process of investigation of the accident.

From the early days of aviation, investigators have turned to Farnborough for expert advice and guidance, and for many years prior to World War II this was supplied largely by Dr W.D. Douglas and his metallurgists. With the advent of the war, and as described in Chapter 2 of this book, a Section was formed of which the author became a founder member.

With domestic rearrangements at Farnborough, the Accidents Section became part of the Airworthiness Department (later Structures), and the Section Head (the author – from December 1957 onwards) became the local co-ordinator of matters concerning accidents and their investigations. The metallurgists have continued to provide a very valuable day-to-day service, not only to the AIB but to many other investigators also. The Accident Section became in effect the Forensic Laboratory and Consultancy for the AIB, particularly where wreckage analysis and technical matters applied. Over the years, that consultancy extended to investigators from all over the world, whether civil or military, a situation unique in aviation, not paralleled elsewhere. Gradually the Section was reduced until the author, still the Head, became virtually the consultant on wreckage analysis at Farnborough.

In latter years, the AIB had its headquarters in London and shared some of the wreckage analysis facilities with the Accident Section at Farnborough, but now it has moved completely to Farnborough and also taken over the facilities. Although at Farnborough, within the confines of the Royal Aircraft Establishment, the AIB remains a completely independent body, as decreed back in 1920.

228

# Index

# Index